VISUALIZING
HUMAN GEOGRAPHY

At Home in a Diverse World

THIRD EDITION

Alyson L. Greiner
Oklahoma State University

VICE PRESIDENT AND EXECUTIVE PUBLISHER Petra Recter	SENIOR CONTENT MANAGER Svetlana Barskaya
EXECUTIVE EDITOR Jessica Fiorillo	SENIOR PRODUCTION EDITOR Elizabeth Swain
ASSOCIATE DEVELOPMENT EDITOR Mallory Fryc	PRODUCT DESIGNER Geraldine Osnato
EDITORIAL ASSISTANT Casseia Lewis	COVER DESIGN Tom Nery
MEDIA SPECIALISTS John Du Val, Laura Byrnes	PHOTO RESEARCHERS Billy Ray, Elizabeth Blomster
SENIOR MARKETING MANAGER Alan Halfen	PRODUCTION SERVICES Aptara

Cover photo credits: Pedestrian traffic (*Rainer Fuhrmann*/Shutterstock); Young African girl (*Lucian Coman*/Shutterstock); Peruvian woman and baby (Gavin *Hellier*/Alamy); Raised hands (*Oleg Troino*/Shutterstock); Arabic businessmen (ESB Professional/Shutterstock); Indian rickshaw (oleander/Shutterstock).

This book was set in ITC New Baskerville Std 10.5 by Aptara and printed and bound by Quad Graphics. The cover was printed by Quad Graphics.

ISBN 978-1-119-32114-9

Printed in the United States of America 10 9 8 7 6 5 4 3 2 1

The inside back cover will contain printing identification and country of origin if omitted from this page. In addition, if the ISBN on the back cover differs from the ISBN on this page, the one on the back cover is correct.

How Is Wiley Visualizing Different?

Wiley Visualizing differs from competing textbooks by uniquely combining three powerful elements: a visual pedagogy, integrated with comprehensive text, the use of authentic situations and issues, and the inclusion of interactive multimedia in the *WileyPLUS* Learning Space learning environment. Together these elements deliver a level of rigor in ways that maximize student learning and involvement. Each key concept and its supporting details have been analyzed and carefully crafted to maximize student learning and engagement.

For more information on the Visualizing series go to http://www.wiley.com//college/visualizing/

Wiley Visualizing and the *WileyPLUS* Learning Space Environment are designed as a natural extension of how we learn

Visuals, comprehensive text, and learning aids are integrated to display facts, concepts, processes, and principles more effectively than words alone can. To understand why the visualizing approach is effective, it is first helpful to understand how we learn.

1. Our brain processes information using two channels: visual and verbal. Our *working memory* holds information that our minds process as we learn. In working memory we begin to make sense of words and pictures and build verbal and visual models of the information.

2. When the verbal and visual models of corresponding information are connected in working memory, we form more comprehensive, or integrated, mental models.

3. After we link these integrated mental models to our prior knowledge, which is stored in our *long-term memory*, we build even stronger mental models. When an integrated mental model is formed and stored in long-term memory, real learning begins.

The effort our brains put forth to make sense of instructional information is called *cognitive load*. There are two kinds of cognitive load: productive cognitive load, such as when we're engaged in learning or exert positive effort to create mental models; and unproductive cognitive load, which occurs when the brain is trying to make sense of needlessly complex content or when information is not presented well. The learning process can be impaired when the amount of information to be processed exceeds the capacity of working memory. Well-designed visuals and text with effective pedagogical guidance can reduce the unproductive cognitive load in our working memory.

How Has Wiley Visualizing Been Shaped by Contributors?

Wiley Visualizing and the *WileyPLUS* learning environment would not have come about without lots of people, each of whom played a part in sharing their research and contributing to this new approach.

Academic Research Consultants

Richard Mayer, Professor of Psychology, UC Santa Barbara. Mayer's *Cognitive Theory of Multimedia Learning* provided the basis on which we designed our program. He continues to provide guidance to our author and editorial teams on how to develop and implement strong, pedagogically effective visuals and use them in the classroom.

Jan L. Plass, Professor of Educational Communication and Technology in the Steinhardt School of Culture, Education, and Human Development at New York University. Plass co-directs the NYU Games for Learning Institute and is the founding director of the CREATE Consortium for Research and Evaluation of Advanced Technology in Education.

Matthew Leavitt, Instructional Design Consultant, advises the Visualizing team on the effective design and use of visuals in instruction and has made virtual and live presentations to university faculty around the country regarding effective design and use of instructional visuals.

Independent Research Studies

SEG Research, an independent research and assessment firm, conducted a national, multisite effectiveness study of students enrolled in entry-level college Psychology and Geology courses. The study was designed to evaluate the effectiveness of Wiley Visualizing. You can view the full research paper at www.wiley.com/college/visualizing/huffman/efficacy.html.

Instructor and Student Contributions

Throughout the process of developing the concept of guided visual pedagogy for Wiley Visualizing, we benefited from the comments and constructive criticism provided by the instructors and colleagues listed below. We offer our sincere appreciation to these individuals for their helpful reviews and general feedback:

Reviewers of Visualizing Human Geography 3e

Geordie Armstrong, *Santa Barbara City College*
Henry Bullamore*, Frostburg State University*
Jacquelyn Chase*, CSU Chico*
Tracy Edwards, *Frostburg State University*

Marilyn Hall, *Carroll Community College*
Taylor Mack, *Louisiana Tech University*
Eileen Pena, *San Jose State University*

Why *Visualizing Human Geography 3e?*

We live in an ever-changing world in which geographical knowledge is central to the well-being of our communities and society. Perhaps nowhere is the urgency of geographical knowledge made clearer to us than through issues involving the local, national, and global impacts of climate change; the British vote to leave the European Union; or the civil war in Syria. Simultaneously, technological innovations continue to open new horizons in mapping and techniques for visualizing geographic information that enable us to see, explore, and understand local and global processes as never before. What a challenging and invigorating time to be either a student or an instructor of geography.

Geographic literacy

Visualizing Human Geography 3e provides an engaging textbook for building geographic literacy and introducing students to the richness of geography, including its many different approaches, perspectives, techniques, and tools. Geographic literacy seeks to endow students with geographic and analytical skills to be creative and capable decision makers and problem solvers. More specifically, geographic literacy includes:

1. fostering the skills of spatial analysis so that students gain an understanding of the importance of scale and can evaluate and interpret the significance of spatial variation;

2. enhancing students' comprehension of the interconnectedness of social and environmental dynamics, and the implications of this for people's livelihoods, their use of the Earth, and environmental change;

3. cultivating global awareness in students and exposing them to divergent views so they are prepared and equipped to participate in an increasingly interconnected world; and

4. educating students about the advantages and limitations of tools such as GIS and GPS in the acquisition and use of geographic information.

A fundamental premise guiding the presentation of material in this book is that such key geographical concepts as place, space, and scale cannot be divorced from a study of process. In other words, questions of why and how are vital to our understanding of where activities, events, or other phenomena are located. Thus, every chapter contains at least one Process Diagram in order to show the diverse factors and complex relations among them that drive social and environmental change.

Human geography is well suited to a visually oriented approach for three reasons. First, maps and images are fundamental tools of geographers that help to reveal patterns or trends that might not otherwise be apparent. Second, within the practice of human geography there is a longstanding tradition of studying cultural landscapes for evidence about such processes as diffusion, urbanization, or globalization in order to more fully understand social difference and to assess human use of the Earth. Third, many human geographers are interested in representation, including the kinds of images that are used by different agencies and entities to characterize places, regions, people, and their activities. Therefore, a visual approach enables a more complete instructional use of photographs, maps, and other visually oriented media to explore and evaluate the significance of different representations.

Other features of this book include:

- content that reflects the latest developments in geographic thought;

- coverage of geographical models and theory as well as their real-world applications;

- top-notch cartography;

- accurate and up-to-date statistics;

- an appendix devoted to understanding map projections.

New to this edition

This Third Edition of *Visualizing Human Geography* offers a new organization as well as new and revised content. In response to reviewer feedback, the order of the chapters has been slightly altered. The chapters "Agricultural Geographies" and "Changing Geographies of Industry and Services" now precede the chapter, "Geographies of Development."

A strong effort has been made to keep the text concise, relevant, and lively. Other changes include:

- **Up-to-date content.** Throughout the text, the information and data have been updated to reflect the most recent data available at the time of the revision.

- **Updated visuals.** Many maps and diagrams have been revised, and a wide variety of new photos have been added throughout to support the learning objectives.

- **New coverage of important topics.** This edition continues the practice of incorporating examples and discussions from relevant current affairs.

- **Chapter 1** (What is Human Geography?) introduces the concept of the Anthropocene to the discussion of human-environment relationships, and features an improved discussion of absolute, relative, and relational space. It also uses a new diagram to link the discussion of GPS, time-space paths, and mobility.

- **Chapter 2** (Globalization and Cultural Geography) incorporates new visuals to explore the geography of multinational corporations, specifically using the example of the Walt Disney Company.

- **Chapter 3** (Population and Migration) now includes a discussion of the relationship between place and quality of life, a new visual depicting the different contributions of developed and developing regions to global population growth, and a new section that examines the relationships between consumption, overpopulation, and the concept of Ecological Footprints.

- **Chapter 4** (Geographies of Language) adds new visuals to the discussion of large languages. Other new visuals and content explore some of the regional patterns in the usage of African American English on Twitter.

- **Chapter 5** (Geographies of Religion) opens with a new feature discussing spirituality and the geography of incense. The chapter also features a new *Process Diagram* that examines the sanctification of the World Trade Center site.

- **Chapter 6** (Geographies of Identity) includes updated discussions of gender roles. These discussions cover the policy changes that now permit women in the military to serve in combat, as well as the controversy over gender-neutral bathrooms. A new *What is Happening in this Picture?* explores the symbolic meanings of jambiyas in Yemen.

- **Chapter 7** (Political Geographies) has a significantly revised section on Global Geopolitics, which includes new coverage of democratic and authoritarian regimes, the geography of freedom, and a case study of the Russian annexation of Crimea. In addition, the material on the European Union now covers the Brexit referendum.

- **Chapter 8** (Urban Geographies) incorporates a new chapter opener on bicycle mobility in Copenhagen. The chapter also now features a new section and supporting visuals related to the foreclosure crisis. A new *Geography InSight* feature focuses on St. Louis, segregation, and events in Ferguson, Missouri.

- **Chapter 9** (Agricultural Geographies) begins with a consideration of our food supply and the value of (and sometimes a distaste for) eating insects. The section on food crises has been revised to emphasize current global food issues such as the consequences of the growing demand for meat.

- **Chapter 10** (Changing Geographies of Industry and Services) adds new visuals to accompany the discussion of commodity dependence, updates the discussion of trends in manufacturing, and incorporates a new *What is Happening in this Picture?* feature on the ship-breakers of Bangladesh.

- **Chapter 11** (Geographies of Development) includes a significantly revised section on global wealth and income inequality, as well as updated information on income inequality in the United States. This is followed by a revised and improved discussion of factors affecting income distribution. A new section discusses alternative approaches to development and covers the United Nations' new agenda for sustainable development.

- **Chapter 12** (Environmental Challenges) presents the latest data and information on patterns of fossil fuel and renewable energy production and consumption. A new discussion of the shale revolution has been added, along with a discussion of the impacts on OPEC. The presentation of land-use and land-cover change has been revised to improve clarity. Coverage of the Paris Agreement has also been added to the material on greenhouse gas reductions.

Instructor Material

(available in *WileyPLUS* Learning Space and on the book companion site)

PowerPoint Presentations

A complete set of highly visual PowerPoint presentations—one per chapter—is available online and in *WileyPLUS* Learning Space to enhance classroom presentations. Tailored to the text's topical coverage and learning objectives, these presentations are designed to convey key text concepts, illustrated by embedded text art. Lecture Launcher PowerPoints also offer embedded links to videos to help introduce classroom discussions with short, engaging video clips.

Test Bank

The Test Bank has a diverse selection of test items including multiple-choice and essay questions that incorporate visuals from the book. The Test Bank is available online in MS Word files and within *WileyPLUS* Learning Space.

Instructor's Manual

The Instructor's Manual includes creative ideas for in-class activities, discussion questions, and lecture transitions.

Clicker PowerPoint Presentations

These PowerPoint Presentations contain relevant questions for each chapter that can be used in class to test the students' knowledge of the material.

Image Gallery

All photographs, figures, maps, and other visuals from the text are online and in *WileyPLUS* Learning Space and can be used as you wish in the classroom. These online electronic files allow you to easily incorporate images into your PowerPoint presentations as you choose, or to create your own handouts.

WILEYPLUS LEARNING SPACE

WileyPLUS Learning Space is an easy way for students to learn, collaborate, and grow. With WileyPLUS Learning Space, students create a personalized study plan, assess progress along the way, and make deeper connections as they interact with the course material and each other.

Through a combination of dynamic course materials and visual reports, this collaborative learning environment gives you and your students immediate insight into strengths and problem areas in order to act on what's most important.

New features for *Visualizing Human Geography* 3e include:

- Interactive Graphics throughout the text that invite the student to participate with the figure in order to learn important pedagogical material.

- Study Tip Videos and Player at the end of each chapter provide valuable information about studying, while the player allows the instructor to customize and add their own videos/content into the course.

WileyPLUS Learning Space

An easy way to help students learn, collaborate, and grow.

Designed to engage today's student, WileyPLUS Learning Space will transform any course into a vibrant, collaborative learning community.

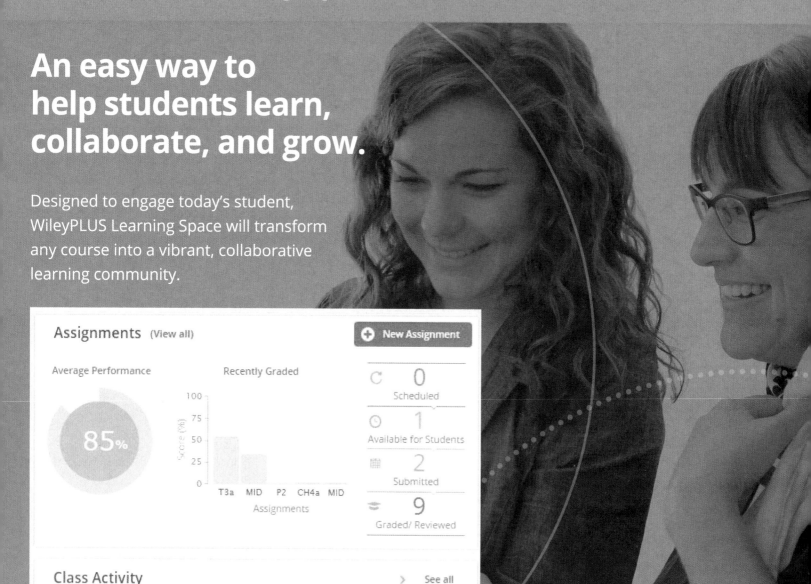

Assignments (View all)

⊕ New Assignment

Average Performance — Recently Graded

85%

Score (%): 100, 75, 50, 25, 0

Assignments: T3a, MID, P2, CH4a, MID

↻ 0 Scheduled

🕓 1 Available for Students

▦ 2 Submitted

≋ 9 Graded/ Reviewed

Class Activity ＞ See all

Identify which students are struggling early in the semester.

Educators assess the real-time engagement and performance of each student to inform teaching decisions. Students always know what they need to work on.

Facilitate student engagement both in and outside of class.

Educators can quickly organize learning activities, manage student collaboration, and customize their course.

Measure outcomes to promote continuous improvement.

With visual reports, it's easy for both students and educators to gauge problem areas and act on what's most important.

www.wileypluslearningspace.com

WILEY

Dedication

For my mother, Annie L. Greiner, who encouraged me to go places.

Special Thanks

This book would not have been possible without the assistance of many talented people. I owe special thanks to Executive Publisher Petra Recter for her support and oversight of this project. From the beginning, Executive Editor Jessica Fiorillo provided unflagging enthusiasm and a keen commitment to geographic education. I am immensely grateful for Associate Development Editor, Mallory Fryc. She not only provided expert guidance along the way, but has been incredibly patient as we have waded through the various challenges that a project of this magnitude presents, not least of which is working with the idiosyncrasies of authors. I also appreciate the efforts of Photo Researchers, Billy Ray and Elizabeth Blomster, who worked diligently to track down elusive photos. I am especially grateful for the adept Production Services provided by Aptara, including the tireless commitment and leadership of Project Manager, Denise Showers. I would like to recognize the expert cartographic assistance provided by the team at Mapping Specialists, Inc. Their work has greatly enhanced the maps in this book.

I also wish to express special thanks to: Casseia Lewis, Editorial Assistant, who helped keep our work running smoothly on a day-to-day basis; John Du Val and Laura Byrnes, Media Specialists, who have been essential to the creation of all manner of high-quality, supporting electronic components; Svetlana Barskaya, Senior Content Manager and Elizabeth Swain, Senior Production Editor, for guiding and overseeing the production of the book; Alan Halfen, Senior Marketing Manager, who has skillfully and steadfastly promoted the book; Geraldine Osnato, Product Designer and Tom Nery, Cover Design, for their vision and creativity.

I greatly appreciate the several reviewers and instructors who provided insightful comments and very helpful suggestions for improving this book. My colleagues in the Department of Geography at Oklahoma State University have likewise been generous in their support for me and this book. I sincerely thank the many students who have used and commented on this book. I remain ever grateful to my husband, Luis D. Montes, for his unwavering support.

About the Author

Alyson L. Greiner is Professor of Geography at Oklahoma State University. She earned her PhD in Geography from the University of Texas at Austin. She has taught courses on cultural geography, world regional geography, the history of geographic thought, and the regional geography of Europe, Africa, and the Pacific Realm. She regularly teaches undergraduate, graduate, and honors students. She has received a Distinguished Teaching Achievement Award from the National Council for Geographic Education. From 2009–2012 she served as a Regional Councilor for the Association of American Geographers. Her scholarly publications include *Anglo-Celtic Australia: Colonial Immigration and Cultural Regionalism* (with Terry G. Jordan-Bychkov) and several peer-reviewed journal articles. She is presently the editor of the *Journal of Cultural Geography*.

Brief Contents

Contents

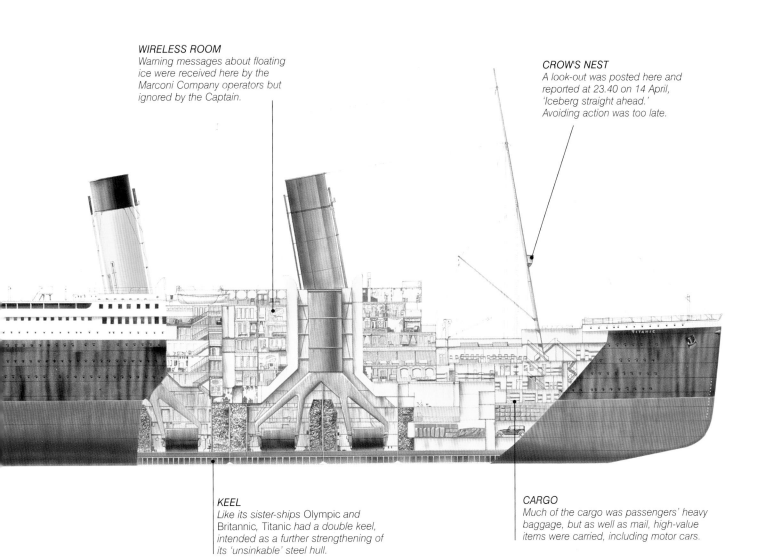

WIRELESS ROOM
Warning messages about floating ice were received here by the Marconi Company operators but ignored by the Captain.

CROW'S NEST
A look-out was posted here and reported at 23.40 on 14 April, 'Iceberg straight ahead.' Avoiding action was too late.

KEEL
Like its sister-ships Olympic and Britannic, Titanic had a double keel, intended as a further strengthening of its 'unsinkable' steel hull.

CARGO
Much of the cargo was passengers' heavy baggage, but as well as mail, high-value items were carried, including motor cars.

Passenger Liners 1900–1929: Part 1

The routes from European ports to New York and Boston were heavily travelled, and competition was intense. Speed was key to success, and the Blue Riband, awarded to the ship making the fastest passage in each direction, was much coveted.

Deutschland

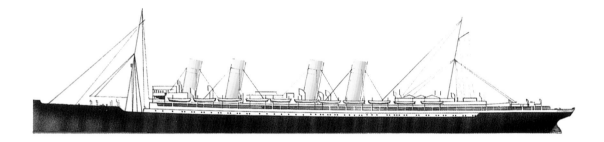

Built to be fastest across the Atlantic, *Deutschland* took the Blue Riband on its maiden voyage and held it for six years. In 1910 it was converted for cruising and renamed *Victoria Luise*. In 1914 it was fitted out as an auxiliary cruiser, though it never served in this role. It was broken up in 1925.

SPECIFICATIONS	
Type:	German liner
Displacement:	16,766 tonnes (16,502 tons)
Dimensions:	208.5m x 20.4m (684ft x 67ft)
Machinery:	Twin screws, quadruple expansion engines
Top speed:	23.6 knots
Route:	Hamburg–New York
Launched:	January 1900

Blücher

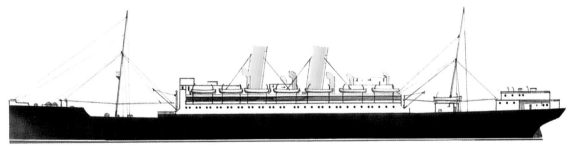

Blücher was built for the Hamburg–Amerika Line and on completion in 1902 was assigned to the Atlantic route. In 1917 it was taken over by the Allies, and in 1919 resumed the New York run under the French flag. Laid up from 1921–23, it was renamed *Suffren* on return to service. It was scrapped in 1929.

SPECIFICATIONS	
Type:	German liner
Displacement:	12,531 tonnes (12,334 tons)
Dimensions:	168m x 19m (549ft 6in x 62ft)
Machinery:	Twin screws, quadruple expansion engines
Top speed:	15 knots
Route:	North Alantic
Launched:	1901

TIMELINE

1900   1901   1903

Baltic

Baltic was built by Harland and Wolff of Belfast for the White Star Line. Capable of carrying nearly 900 passengers plus 2000 steerage, it was the world's largest ship when launched. Used as a troop transport during World War I, it was re-boilered in 1924 and laid up in 1932. It was broken up in 1933.

SPECIFICATIONS

Type:	British merchant vessel
Displacement:	24,258 tonnes (23,876 tons)
Dimensions:	221m x 23m (725ft x 75ft 6in)
Machinery:	Twin screws, triple expansion engines
Top speed:	17 knots
Cargo:	General cargo
Route:	North Atlantic
Launched:	November 1903

Empress of Britain

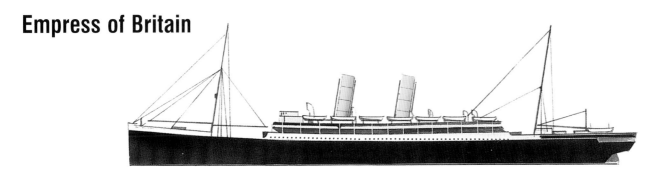

Entering service for the Canadian Pacific Company in 1906, *Empress of Britain* carried 1460 passengers in three classes. On the outbreak of World War I, it was taken over as an auxiliary cruiser and later served as a troop transport. In 1924, it underwent a major refit, was renamed *Montroyal*, and was scrapped in 1930.

SPECIFICATIONS

Type:	Canadian liner
Displacement:	14,416 tonnes (14,189 tons)
Dimensions:	167m x 20m (549ft x 66ft)
Machinery:	Twin screws, quadruple expansion engines
Top speed:	20 knots
Route:	Liverpool–Canada
Launched:	November 1905

Lusitania

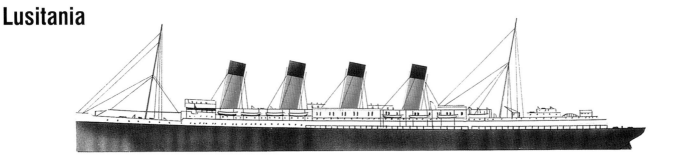

At launch, *Lusitania*, built for Cunard's North Atlantic service, was the largest ship in the world, carrying 2165 passengers – and the Blue Riband holder in both directions in 1907. Returning from New York to Southampton in May 1915, it was sunk off Ireland by the German submarine *U-20* for the loss of 1198 lives.

SPECIFICATIONS

Type:	British liner
Displacement:	32,054 tonnes (31,550 tons)
Dimensions:	232m x 27m (761ft x 88ft)
Machinery:	Quadruple screws, turbines
Top speed:	24 knots
Route:	North Atlantic
Launched:	1906

1905

1906

Passenger Liners 1900–1929: Part 2

Despite an accent on luxury, all steam liner companies knew that most passengers travelled third class. They were segregated from first class but were still a cut above the bottom-rung 'steerage' passengers, mostly emigrants going one way only.

France

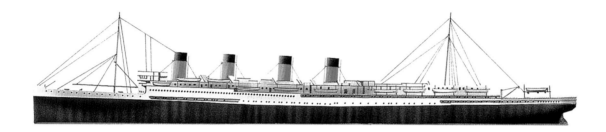

France was one of the smallest transatlantic liners, but also one of the fastest. In World War I, it was an auxiliary cruiser, troop transport and hospital ship. In 1919, it returned to the Atlantic route, so popular that passengers had to bid for their cabins. After a last crossing in 1932 it was broken up in 1934/35.

SPECIFICATIONS	
Type:	French liner
Displacement:	27,188 tonnes (26,760 tons)
Dimensions:	217.2m x 23m (712ft 7in x 75ft 6in)
Machinery:	Four screws, turbines
Top speed:	25.9 knots
Route:	Le Havre–New York
Launched:	September 1910

Franconia

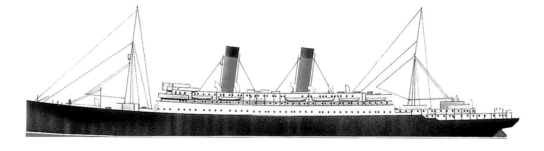

Similar in appearance to the other Cunard ships Laconia and Caronia, and used chiefly on the Liverpool–Boston service, Franconia also specialized in winter cruises. Passenger capacity was 300 first, 350 second and 2200 third class. A troopship during World War I, it was sunk by U-boat torpedoes in 1916.

SPECIFICATIONS	
Type:	British passenger liner
Displacement:	18,441 tonnes (18,150 tons)
Dimensions:	182.95m x 21.75m (600ft 3in x 71ft 3in)
Machinery:	Twin screws, quadruple expansion engines
Top speed:	17 knots
Route:	North Atlantic, Mediterranean cruising
Launched:	1910

TIMELINE

1910   1911

Kaiser Franz Josef I

SPECIFICATIONS	
Type:	Austrian liner
Displacement:	17,170 tonnes (16,900 tons)
Dimensions:	152.4m x 18.8m x 8.8m (500ft x 62ft x 29ft)
Machinery:	Twin screws, quadruple expansion engines
Top speed:	19 knots
Routes:	Trieste–New York, Trieste–Buenos Aires
Launched:	September 1911

During World War I, *Kaiser Franz Josef I* was laid up at Trieste. In 1919 it was handed to Italy and renamed *Presidente Wilson*. From 1920 to 1922, it was Italy's largest liner. Transferred to the Lloyd Triestino and then Adriatica lines, it worked through the 1930s, and was scuttled at La Spezia in 1944.

Matsonia

SPECIFICATIONS	
Type:	US cargo/passenger liner
Displacement:	9886.5 tonnes (9728 tons)
Dimensions:	146m x 17.5m x 9m (480ft 6in x 58ft x 30ft 6in)
Machinery:	Single screw, triple expansion engine
Top speed:	15 knots
Cargo:	Tropical fruit, sugar, general merchandise
Route:	San Francisco–Hawaii
Launched:	1913

Matsonia was built for Matson Lines, its machinery and funnel set unusually far back. Passenger accommodation in two classes was 329. During the 1932–37 slump, it was laid up. In both World Wars, it was used by the US government as an armed merchant cruiser. It was broken up in 1957.

Britannic

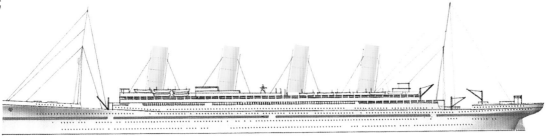

SPECIFICATIONS	
Type:	British hospital ship
Displacement:	48,928 tonnes (48,158 tons)
Dimensions:	275m x 27m (903ft x 94ft)
Machinery:	Triple screws, geared turbines
Top speed:	21 knots
Launched:	1914

Britannic was the largest of a trio of giant liners ordered by the White Star Line from Harland and Wolff, the others being *Olympic* and *Titanic*. In 1915, the Admiralty ordered its completion as a hospital ship. In 1916, it struck a German mine in the Aegean Sea, and an hour later it capsized and sank.

1913   1914

Passenger Liners 1900–1929: Part 3

Warfare emptied the passenger liners, though some companies leased their vessels as troopships or hospital ships. Many liners were laid up for the duration of World War I, and not recommissioned afterwards, being scrapped before their time.

Duilio

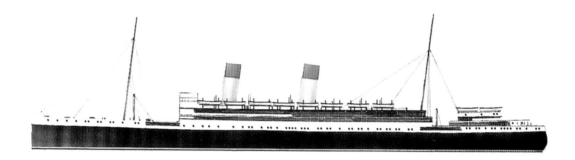

Italy's first home-built liner to exceed 20,320 tonnes (20,000 tons), *Duilio* was completed in 1923 for the North American route. In 1928 it switched to the South American route, and in 1933 began a South African service. It was chartered by the International Red Cross in 1942. Sunk in 1944, it was raised for scrap in 1948.

SPECIFICATIONS

Type:	Italian liner
Displacement:	24,670 tonnes (24,281 tons)
Dimensions:	193.5m x 23.2m (634ft 10in x 76ft)
Machinery:	Quadruple screws, turbines
Routes:	Genoa–New York; Genoa–Buenos Aires
Launched:	1916

Columbus

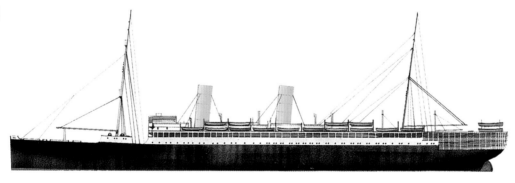

Laid down in 1914 for Norddeutscher Lloyd, *Columbus*'s building was halted until 1920. Passenger capacity was 1837 in three classes and it was a popular ship. In 1929 new turbines were installed, increasing its speed. It was scuttled in the Atlantic in December 1939 to prevent capture by the British HMS *Hyperion*.

SPECIFICATIONS

Type:	German liner
Displacement:	32,871 tonnes (32,354 tons)
Dimensions:	236m x 25m (775ft x 83ft)
Machinery:	Twin screws, triple expansion engine, replaced by turbines
Top speed:	19, later 23 knots
Route:	North Atlantic, Germany–United States
Launched:	August 1922

TIMELINE

1916   1922

Doric

The White Star Line's only turbine-driven liner, *Doric* was built for the Liverpool–Canada route, carrying 583 cabin class and 1688 third-class passengers. By 1930, it took 320 cabin class, 657 tourist class and 537 third-class passengers. Damaged in 1935 by a collision with the freighter *Formigny*, it was sold for scrap.

SPECIFICATIONS	
Type:	British liner
Displacement:	28,935 tonnes (28,480 tons)
Dimensions:	183m x 20.6m (600ft 6in x 67ft 6in)
Machinery:	Twin screws, turbines
Top speed:	16 knots
Route:	Liverpool–Halifax, Montreal
Launched:	1922

Eridan

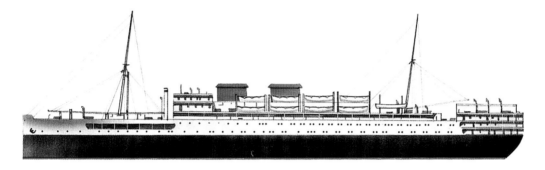

The largest and most powerful motorship yet built in France, *Eridan* was built for the Australian service. During World War II, it was under Vichy control until captured in 1942. After a refit in 1947, *Eridan* served on the Indian route. In another refit in 1951, it acquired a large single funnel. It was broken up in 1956.

SPECIFICATIONS	
Type:	French liner
Displacement:	14,361 tonnes (14,135 tons)
Dimensions:	142.6m x 18.5m (468ft 7in x 61ft)
Machinery:	Twin screws, diesel engines
Top speed:	16 knots
Routes:	Marseille–New Caledonia; Marseille–India
Launched:	June 1928

Europa

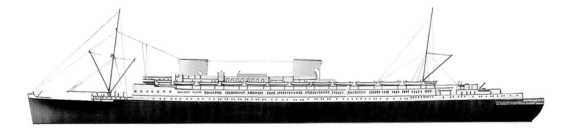

With an unusual raked bow and bulbous forefoot designed for easier passage through the water, *Europa* was completed in 1930. In 1946 it became the French *Liberté*, but while laid up, broke away and collided with the sunken wreck of the liner *Paris* and sank. Resuming service in 1950, it was broken up in 1962.

SPECIFICATIONS	
Type:	German liner
Displacement:	50,542 tonnes (49,746 tons)
Dimensions:	285m x 31m (930ft 9in x 102ft 1in)
Machinery:	Quadruple screws, turbines
Top speed:	27.9 knots
Route:	Bremen–Le Havre–New York
Launched:	July 1928

1928

Lake Steamers & River Vessels

Lake and river ships floated on fresh water, which is less buoyant than salt –
something that designers had to bear in mind for bigger vessels. And some lake
ships were large, particularly on the Great Lakes of North America. Most were
modest-sized, like the steamers on Lake Titicaca and the East African Lakes.

Baikal

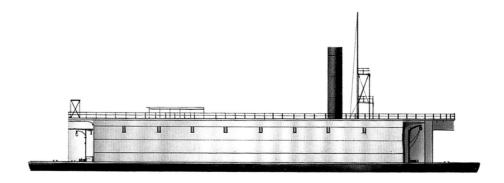

Baikal was built to connect the eastern and western sections of the Trans-
Siberian railway across Lake Baikal, and was also used as an ice-breaker. Built
on the Tyne in the 1890s, it was dismantled for transportation, and reassembled.
It served on the lake until the railway was finally extended round the shore.

SPECIFICATIONS	
Type:	Russian train ferry
Displacement:	2844 tonnes (2800 tons)
Dimensions:	76.2m x 19.2m (250ft x 63ft)
Machinery:	Twin screws, vertical triple expansion engines
Launched:	June 1900

Lady Hopetoun

Built in Sydney as an inspection boat for the officials of the Maritime Services
Board, *Lady Hopetoun* worked in Sydney Harbour. It is a typical example of the
steam pinnace, once a common sight. Most have vanished, but this one has
been preserved and has been a museum craft in Sydney since 1991.

SPECIFICATIONS	
Type:	Australian harbour inspection boat
Displacement:	38.6 tonnes (38 tons)
Dimensions:	23.45m x 4.2m x 2.05m (70ft x 13ft 9in x 6ft 9in)
Machinery:	Single screw, triple expansion engine
Launched:	1902

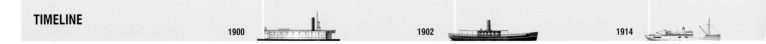

TIMELINE

1900

1902

1914

Liemba

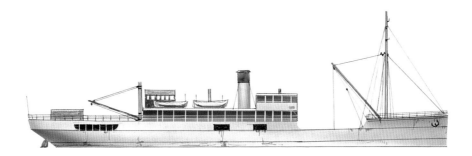

SPECIFICATIONS	
Type:	East African lake steamer
Displacement:	1600 tonnes (1575 tons)
Dimensions:	70.7m x 10.05m x 2.75m (232ft x 33ft x 9ft)
Machinery:	Twin screws, triple expansion engine
Top speed:	10 knots
Route:	Lake Tanganyika, Kigoma–Mpulungu
Launched:	1914

Built in sections in Germany and assembled at Kigoma (then a German colony), this ship was first named *Graf von Goetzen*. In 1916 it was scuttled to prevent the British taking it. Raised in 1924, it was refurbished, renamed *Liemba* and restored to service, carrying 384 passengers and light freight between railheads.

Oscar Huber

SPECIFICATIONS	
Type:	German river tug
Displacement:	203.2 tonnes (200 tons)
Dimensions:	75m x 20.7m x 1.55m (246ft x 67ft 10in x 5ft 1in)
Machinery:	Sidewheels, triple expansion engine
Route:	River Rhine
Launched:	1922

Steam tugs of this type towed long strings of barges on the Rhine and its linked waterways from central Germany to the coast. Sidewheels enabled them to have a shallow draught that could cope with reduced river flows. *Oscar Huber* is maintained as a museum craft at Duisburg, where it was built.

William G. Mather

SPECIFICATIONS	
Type:	US Great Lakes bulk carrier
Displacement:	8800 tonnes (8662 tons)
Dimensions:	183.2m x 18.9m x 5.5m (601ft x 62ft x 18ft)
Machinery:	Single screw, geared turbine engine
Top speed:	12 knots
Cargo:	Iron ore, wheat
Route:	Great Lakes
Launched:	1925

For navigating the Great Lakes and their linking canals, a unique ship design was produced, with command position and engines at opposite ends. The deck was free of equipment, since loading and unloading gear was available at both ends of the voyage. This example from the 1920s is retained as a museum ship.

1922   1925

Specialized Vessels 1900–1929

Some ships were built for a specific purpose that required either a particular type of hull, perhaps ice-resistant or shallow-draught, or fitting with specialized equipment. Most common were tug-boats, from small harbour tugs to ocean-going salvage vessels, and the various kinds of dredger, but the range of requirements was wide.

Discovery

Built in Dundee, Scotland, and designed for Polar research work, *Discovery* had a wooden hull reinforced to withstand ice pressure. In 1901, it took Captain Scott to the Antarctic. Also used on numerous other research voyages, it was berthed on the Thames for many years and is now a museum ship in Dundee.

SPECIFICATIONS

Type:	British exploration ship
Displacement:	1646 tonnes (1620 tons)
Dimensions:	52m x 10m x 4.8m (172ft x 34ft x 15ft 8in)
Machinery:	Single screw, triple expansion engine
Rigging:	Three masts, barque rig
Top speed:	8 knots
Launched:	March 1901

Industry

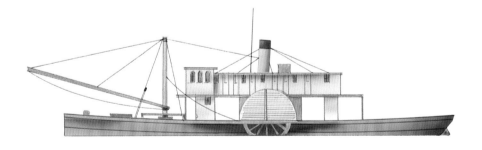

Industry's job was to keep the Murray's navigable channel free of obstructions such as floating logs and to act as a mobile workshop for maintenance work on locks and wharves. Its derrick could operate a bucket dredge to clear silt. In regular use until 1969, it is now a museum vessel at Renmark, South Australia.

SPECIFICATIONS

Type:	Australian dredger and snag boat
Displacement:	92.4 tonnes (91 tons)
Dimensions:	34.15m x 5.65m x .94m (112ft x 18ft 6 in x 3ft 1in)
Machinery:	Sidewheels, 30hp steam engine
Route:	Murray River
Launched:	1911

TIMELINE

 1901    1911    1913

Acadia

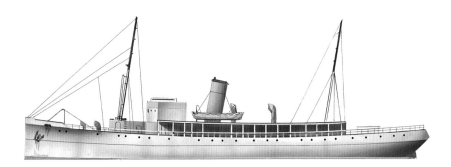

Interest in Canada's northlands was strong around 1910 and *Acadia* was a specialized and well-equipped hydrographic research ship intended to provide valuable knowledge of the coastline and inshore waters, and to underline Canada's claim to the region. *Acadia* is preserved at Halifax, Nova Scotia.

SPECIFICATIONS	
Type:	Canadian hydrographic survey ship
Displacement:	859.5 tonnes (846 tons)
Dimensions:	51.8m x 10.25m x 3.65m (170ft x 33ft 6in x 12ft)
Machinery:	Single screw, vertical triple expansion engine
Top speed:	12.5 knots
Route:	Canadian east and north-east coasts
Launched:	1913

Inverlago

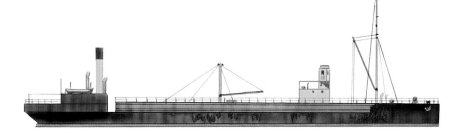

Serving the Venezuelan oilfield, ships like *Inverlago* were of restricted depth because of the shallow waters of the lake. Even so, it could carry 3200 tonnes (3156 tons) of crude oil from the well-heads to deep-water berths on the sea-coast. By 1953, the channels had been deepened and ships of such designs were not needed.

SPECIFICATIONS	
Type:	Dutch–Venezuelan oil tanker
Displacement:	2758.6 tonnes (2600 tons)
Dimensions:	92.5m x 11.6m x 4.05m (305ft x 38ft x 13ft 3in)
Machinery:	Single screw, triple expansion engine
Top speed:	10 knots
Cargo:	Crude oil
Route:	Lake Maracaibo
Launched:	1925

Artiglio II

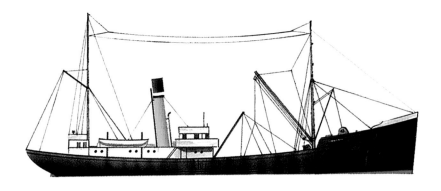

Artiglio II was a small coaster adapted in 1929 to recover gold from the liner *Egypt*, which had sunk in 1922. The exercise marked a new era in underwater salvage, as the liner had gone down in a depth of 110m (360ft) – too deep for conventional diving gear. Most of the gold, valued at £1,054,000, was recovered.

SPECIFICATIONS	
Type:	Italian salvage vessel
Displacement:	305 tonnes (300 tons) (approx)
Dimensions:	42.6m x 7.6m x 2.1m (139ft 9in x 25ft x 7ft) (approx)
Top speed:	14 knots (approx)
Converted:	1929

1925

1929

SHIP TYPES 1930–1949

The economic depression of the 1930s slowed down the rate of construction of new ships, but technical development was intensified in the effort to get greater economy and efficiency.

Oil fuel replaced coal in all new vessels, and the diesel engine brought about the 'motor ship' to compete with the steamer. Attempts to agree on limits to warship sizes collapsed with the imminence of war in 1939. In World War II, more than twenty million tons of shipping were sunk, with huge loss of life, and yet most of it was speedily replaced.

Left: *Bismarck*, with *Tirpitz*, was the largest German warship ever built. Its one and only combat mission became a naval epic of World War II.

Aircraft Carriers (Escort)

Large carriers took time to design and build, and from early in World War II, the hulls of existing ships were requisitioned and converted into small or medium-sized carriers in order to help provide air cover for battleships and convoys. Their facilities were usually limited, but they provided vital support at a crucial time.

Audacity

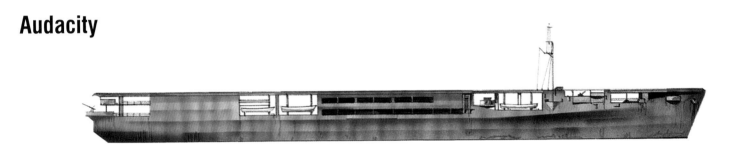

The captured German merchant ship *Hannover* was given a 140m x 18m (400ft x 60ft) flight deck above the hull. There was no hangar or elevator, so its aircraft had to remain on the deck, exposed to the weather. As *Audacity*, it served for only six months before it was torpedoed and sunk by *U-751* in December 1941.

SPECIFICATIONS	
Type:	British escort carrier
Displacement:	11,176 tonnes (11,000 tons)
Dimensions:	142.4m x 17.4m x 7.5m (467ft 5in x 57ft x 24ft 6in)
Machinery:	Single screw, diesel engine
Top speed:	15 knots
Main armament:	One 102mm (4in) gun, eight AA guns
Aircraft:	Six Martlet fighters
Complement:	480
Launched:	1939 (converted 1941)

Sangamon

First a commercial tanker, then a fleet oiler, *Sangamon* was one of four similar ships converted to carriers. The aft-placed engines allowed for considerable hangar space. The wooden flight deck was fitted with two lifts and a catapult. The ships could carry 12,000 tonnes of oil, enabling them to act as refuellers.

SPECIFICATIONS	
Type:	US escort carrier
Displacement:	24,257 tonnes (23,875 tons)
Dimensions:	168.55m x 32.05m x 9.3m (553ft x 105ft 2in x 30ft 7in)
Machinery:	Twin screws, geared turbines
Top speed:	18 knots
Main armament:	Two 127mm (5in) guns
Aircraft:	30 (later 36)
Complement:	1080
Launched:	1939 (converted 1943)

TIMELINE   1939   1940

Dixmude

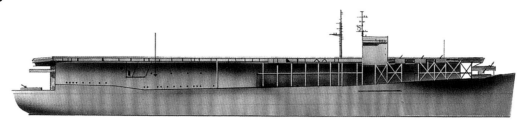

This ship was built in the United States as *Rio Parana* for lease to Britain. On arrival, the flight deck was increased to 134m (440ft) and the ship became HMS *Biter*. Passed to France in 1945, it was renamed *Dixmude*, serving as an aircraft transport. Disarmed as an accommodation ship, it was scrapped in 1966.

SPECIFICATIONS	
Type:	French aircraft carrier
Displacement:	11,989 tonnes (11,800 tons)
Dimensions:	150m x 23m x 7.6m (490ft 10in x 78ft x 25ft 2in)
Machinery:	Single screw, diesel engine
Top speed:	16.5 knots
Main armament:	Three 102mm (4in) guns
Aircraft:	15
Launched:	December 1940

Activity

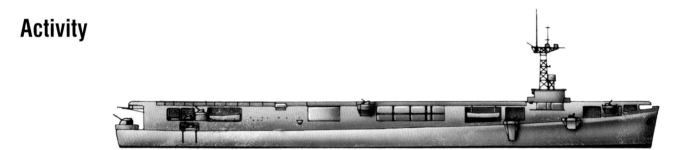

Built on the hull intended for a refrigerated cargo ship, *Activity* served as a carrier but was compromised by its restricted hangar space, only 31m (100ft long). But it was an effective escort to Arctic and Atlantic convoys and carried aircraft to the Far East. After 1945, it was reconstructed as the cargo ship *Breconshire*.

SPECIFICATIONS	
Type:	British escort carrier
Displacement:	14,529 tonnes (14,300 tons)
Dimensions:	156.3m x 20.3m x 7.95m (512ft 9in x 66ft 6in x 26ft)
Machinery:	Twin screws, diesels
Top speed:	18 knots
Main armament:	Two 102mm (4in) guns, twenty-four 20mm (0.78in) AA guns
Aircraft:	10
Complement:	700
Launched:	1942

Dédalo

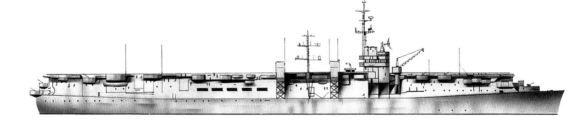

Dédalo began existence as the US carrier *Cabot*, completed in 1943. After service in World War II, it was decommissioned in 1947. Laid up for 20 years, it was lent to Spain in 1967, and purchased outright in 1972. It remained in service until the new carrier *Principe de Asturias* was commissioned in 1982.

SPECIFICATIONS	
Type:	Spanish aircraft carrier
Displacement:	16,678 tonnes (16,416 tons)
Dimensions:	190m x 22m x 8m (622ft 4in x 73ft x 26ft)
Machinery:	Quadruple screws, turbines
Top speed:	30 knots
Main armament:	Twenty-six 40mm (1.6in) guns
Aircraft:	20
Launched:	April 1943

1942   1943

Enterprise

Commissioned in May 1938, *Enterprise* was engaged in almost every major carrier battle of World War II. At the Battle of Midway, its dive bombers helped sink four Japanese carriers. A refit was carried out in late 1943. In 1947, it was placed on the reserve list and it went for breaking in 1958.

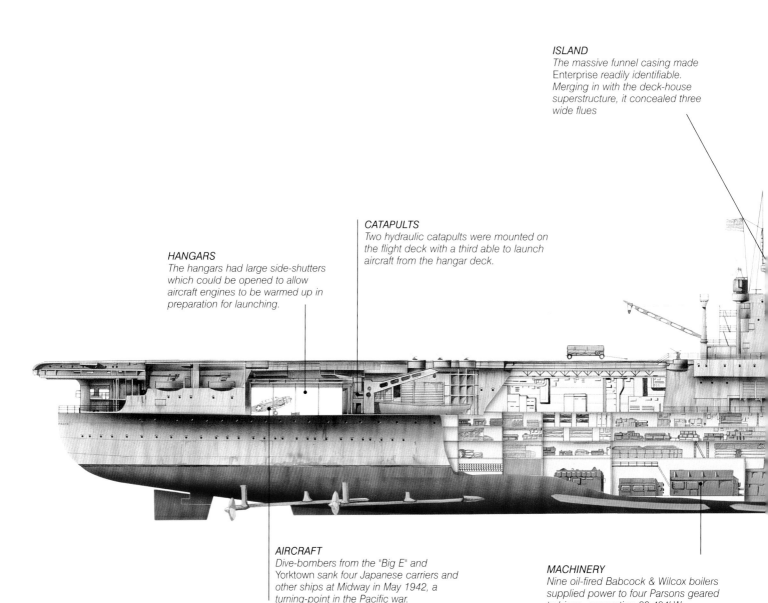

ISLAND
The massive funnel casing made Enterprise readily identifiable. Merging in with the deck-house superstructure, it concealed three wide flues

CATAPULTS
Two hydraulic catapults were mounted on the flight deck with a third able to launch aircraft from the hangar deck.

HANGARS
The hangars had large side-shutters which could be opened to allow aircraft engines to be warmed up in preparation for launching.

AIRCRAFT
Dive-bombers from the "Big E" and Yorktown sank four Japanese carriers and other ships at Midway in May 1942, a turning-point in the Pacific war.

MACHINERY
Nine oil-fired Babcock & Wilcox boilers supplied power to four Parsons geared turbines, generating 89,484kW (120,000 shp).

Enterprise

Enterprise was involved in Midway, Guadalcanal, the Eastern Solomons, the Gilbert Islands, Kwajalein, Eniwetok, the Truk raid, Hollandia, Saipan, the Battle of the Philippine Sea, Palau, Leyte, Luzon, Taiwan, the China coast, Iwo Jima and Okinawa. She received five bomb hits and survived two attacks by kamikazes off Okinawa.

SPECIFICATIONS	
Type:	US aircraft carrier
Displacement:	25,908 tonnes (25,500 tons)
Dimensions:	246.7m x 26.2m x 7.9m (809ft 6in x 86ft x 26ft)
Machinery:	Quadruple screws, turbines
Top speed:	37.5 knots
Main armament:	Eight 127mm (5in) guns
Aircraft:	96
Complement:	2175
Launched:	October 1936

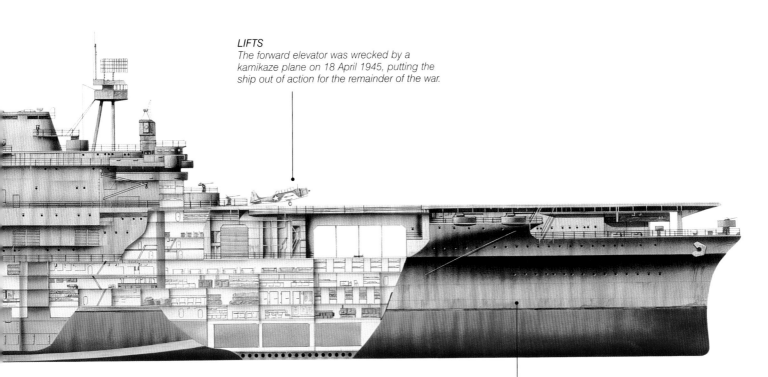

LIFTS
The forward elevator was wrecked by a kamikaze plane on 18 April 1945, putting the ship out of action for the remainder of the war.

HULL
Though purpose-built as a carrier, the hull design clearly shows how the form evolved from the cruiser hull.

American Carriers

On entering World War II, the United States had few aircraft carriers, but began to construct them on an enormous scale, with many auxiliary and escort carriers built on merchant ship hulls. Carrier-borne aircraft played a vital role in the great battles of the Pacific Ocean, and became the prime destroyers of submarines.

Wasp

Notable among purpose-built US carriers for its lack of armour protection, *Wasp* was commissioned in April 1940. It helped in the defence of Malta by delivering 100 Spitfires. Transferred to the Pacific in June 1942, it was torpedoed by the Japanese submarine *I19* off Guadalcanal on 15 September, and abandoned.

SPECIFICATIONS	
Type:	US aircraft carrier
Displacement:	18,745.32 tonnes (18,450 tons)
Dimensions:	225.9m x 28.35m x 7.45m (741ft 3in x 93ft x 24ft 6in)
Machinery:	Twin screws, geared turbines
Top speed:	29.5 knots
Aircraft:	76
Main armament:	Eight 127mm (5in) guns
Armour:	37mm (1.5in) deck
Complement:	2167
Launched:	1939

Essex

The needs of the navy for air cover led to great increases in the size of aircraft carriers, with larger hulls to stow the fuel required for up to 91 aircraft. There were 24 vessels in the *Essex* class. *Essex* entered service in 1942, was removed from the effective list in 1969 and scrapped in 1973.

SPECIFICATIONS	
Type:	US aircraft carrier
Displacement:	35,438 tonnes (34,880 tons)
Dimensions:	265.7m x 29.2m x 8.3m (871ft 9in x 96ft x 27ft 6in)
Machinery:	Quadruple screws, turbines
Top speed:	32.7 knots
Aircraft:	90–100
Main armament:	Twelve 127mm (5in) guns
Launched:	July 1942

TIMELINE

1939   1942

Independence

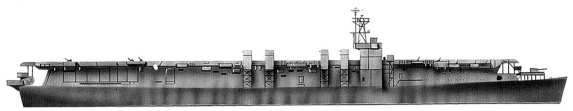

Borrowing the frame of a *Cleveland*-class light cruiser, *Independence* was part of an emergency carrier programme. Nine vessels entered service in 1943. *Independence* had room to ferry up to 100 aircraft. It was used as a target in the Bikini atomic bomb tests, and was sunk as a target in 1951.

SPECIFICATIONS	
Type:	US aircraft carrier
Displacement:	13,208 tonnes (13,000 tons)
Dimensions:	190m x 33m x 7.6m (623ft x 109ft 3in x 25ft 11in)
Machinery:	Quadruple screws, turbines
Top speed:	31 knots
Main armament:	Two 127mm (5in) guns
Armour:	140mm (5.5ins) belt, 51mm (2in) deck
Aircraft:	45
Launched:	August 1942

Gambier Bay

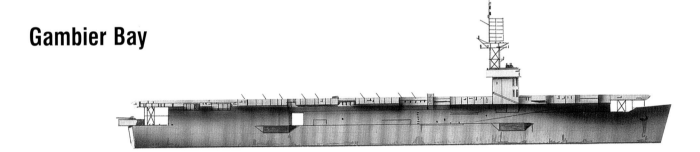

Gambier Bay was another of the 50-strong group of light escort carriers assembled on merchant ship hulls. All were built in under a year. In early 1944, *Gambier Bay* ferried aircraft to USS *Enterprise* and supported action off Saipan, in the Marianas and at Leyte. It was sunk by gunfire off Samar in October 1944.

SPECIFICATIONS	
Type:	US escort carrier
Displacement:	11,074 tonnes (10,900 tons)
Dimensions:	156.1m x 32.9m x 6.3m (512ft 3in x 108ft x 20ft 9in)
Machinery:	Twin screws, reciprocating engines
Top speed:	19 knots
Aircraft:	28
Launched:	November 1943

Attu

In 1942, shipbuilder Henry J. Kaiser was mass producing cargo vessels to replace those lost in war. With a severe shortage of aircraft carriers, the decision was taken to adapt 50 of the unfinished hulls as escort carriers. *Attu* was built in 75 days. It served in the Pacific until 1946, and was scrapped in 1949.

SPECIFICATIONS	
Type:	US escort carrier
Displacement:	11,076 tonnes (10,902 tons)
Dimensions:	156.1m x 32.9m x 6.3m (512ft 3in x 108ft x 20ft 9in)
Machinery:	Twin screws, reciprocating engines
Top speed:	19.3 knots
Main armament:	One 127mm (5in) gun
Launched:	1944

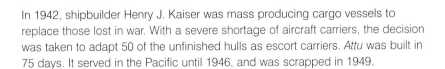

1943  1944

British Carriers

In 1939, the British Navy had the advantage of twenty years' experience in carrier design and operation. But their number was not large. In 1938, a construction programme was got under way, and war demands and conditions prompted further designs in 1940. By then, the carriers were in the thick of naval warfare.

Ark Royal

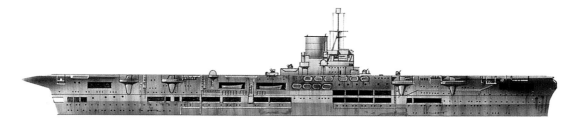

Ark Royal was the Royal Navy's first large purpose-built aircraft carrier, with a long flight deck some 18m (60ft) above the deep water load line. Its full complement was 60 aircraft, although it never carried this many. Part of Force H in the Mediterranean, *Ark Royal* was sunk in November 1941 by submarine *U-81*.

SPECIFICATIONS	
Type:	British aircraft carrier
Displacement:	28,164 tonnes (27,720 tons)
Dimensions:	243.8m x 28.9m x 8.5m (800ft x 94ft 9in x 27ft 9in)
Machinery:	Triple screws, geared turbines
Top speed:	31 knots
Main armament:	Sixteen 114mm (4.5in) guns
Armour:	114mm (4.5in) belt, 7.6mm (3in) bulkheads
Aircraft:	50–60
Launched:	April 1937

Formidable

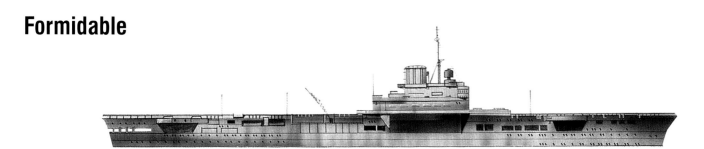

Formidable was modelled on *Ark Royal* but with one hangar deck. As the nature of the coming war was becoming clear, attention was given to its armour and defensive capacity; it withstood two kamikaze attacks in May 1945. The flight deck was widened to carry extra planes. *Formidable* was scrapped in 1953.

SPECIFICATIONS	
Type:	British aircraft carrier
Displacement:	28,661 tonnes (28,210 tons)
Dimensions:	226.7m x 29.1m x 8.5m (743ft 9in x 95ft 9in x 28ft)
Machinery:	Triple screws, turbines
Top speed:	30.5 knots
Main armament:	Sixteen 114mm (4.5in) guns
Armour:	115mm (4.5in) hangars, 76mm (3in) deck
Aircraft:	36, later 54
Complement:	1229, later 1997
Launched:	August 1939

TIMELINE

1937 1939 1940

Indomitable

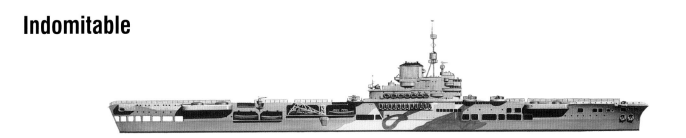

Indomitable was modified from the *Illustrious* class carriers, its armour weight sacrificed to provide more aircraft space, with an additional hangar deck. *Indomitable* was damaged by bombs in August 1942 and by an aerial torpedo in August 1943. Back in service from April 1944, it was broken up in 1955.

SPECIFICATIONS	
Type:	British aircraft carrier
Displacement:	30,205 tonnes (29,730 tons)
Dimensions:	229.8m x 29.2m x 8.85m (753ft 11in x 95ft 9in x 29ft)
Machinery:	Triple screws, geared turbines
Top speed:	30.5 knots
Main Armament:	16 114mm (4.5in) guns
Armour:	114mm (4.5in) belt, 37mm (1.5in) hangar sides, 76mm (3in) flight deck
Aircraft:	45 (later 56)
Complement:	1392 (later 2100)
Launched:	1940

Unicorn

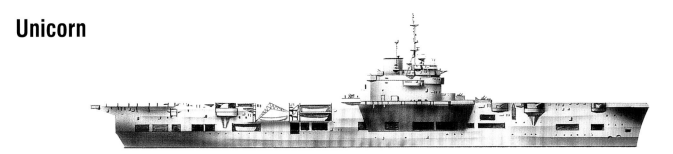

Unicorn was meant to be a depot/maintenance support ship. It was modified during construction to operate its own aircraft, as well as repair those from other carriers. During World War II, it served in the Mediterranean, Atlantic and Pacific. It later became a depot ship in Hong Kong, and was scrapped in 1959/60.

SPECIFICATIONS	
Type:	British aircraft carrier
Displacement:	20,624 tonnes (20,300 tons)
Dimensions:	186m x 27.4m x 7.3m (610ft x 90ft x 24ft)
Machinery:	Twin screws, turbines
Top speed:	24 knots
Main armament:	Eight 102mm (4in) guns
Armour:	51mm (2in) flight deck, 76–51mm (3–2in) on magazines
Aircraft:	36
Complement:	1200
Launched:	November 1941

Eagle

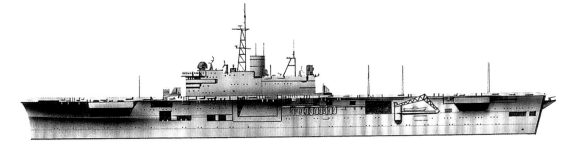

During construction of the *Illustrious* class of 1938, designs were prepared in 1942 for their successors. These allowed for two complete hangars and ability to handle the heavier aircraft due to be introduced. *Eagle* entered service in 1951, was decommissioned in 1972 and was sent for breaking-up in 1978.

SPECIFICATIONS	
Type:	British aircraft carrier
Displacement:	47,200 tonnes (46,452 tons)
Dimensions:	245m x 34m x 11m (803ft 9in x 112ft 9in x 36ft)
Machinery:	Quadruple screws, turbines
Top speed:	31 knots
Main armament:	Sixteen 114mm (4.5in) guns; six GSW Seacat SAM (from 1962)
Aircraft:	80
Complement:	2750 including air group
Launched:	March 1946

1941

1946

German & Japanese Aircraft Carriers

The Japanese Navy, an early user of carriers, built several in the 1930s and also ships easily convertible to carriers. Its carrier-borne aircraft attacked Pearl Harbor in December 1941. Germany, however, had been prevented from building carriers in the 1920s, and its designers and yards had no experience of this type of vessel.

Ryujo

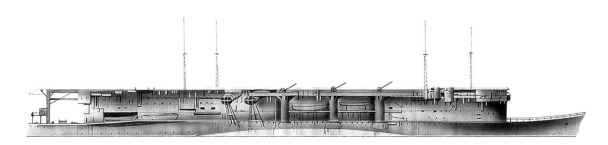

Japan's first major purpose-built aircraft carrier was designed with a cruiser hull, restricting its width. A second hangar above the first gave excessive top weight, requiring post-launch modification. The hull was strengthened between 1934 and 1936, and the bulges widened. Aircraft from USS *Saratoga* sank it in 1942.

SPECIFICATIONS	
Type:	Japanese aircraft carrier
Displacement:	10,150 tonnes (9990 tons)
Dimensions:	175.3m x 23m x 5.5m (575ft 5in x 75ft 6in x 18ft 3in)
Machinery:	Twin screws, turbines
Top speed:	29 knots
Main armament:	Twelve 127mm (5in) guns
Aircraft:	48
Complement:	600
Launched:	April 1931

Zuiho

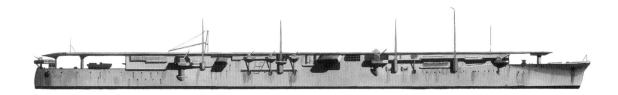

Many Japanese carriers had minimal deck superstructure. *Zuiho*, a submarine tender that was later converted, had none. Nor did it have armour protection. Damaged at Santa Cruz, it was repaired, and saw action at the Marianas and Guadalcanal. It was sunk at Cape Engano, in the Battle of Leyte Gulf.

SPECIFICATIONS	
Type:	Japanese aircraft carrier
Displacement:	14,528.8 tonnes (14,300 tons) full load
Dimensions:	204.8m x 18.2m x 6.65m (332ft x 59ft 89n x 21ft 9in)
Machinery:	Twin screws, geared turbines
Top speed:	28 knots
Main armament:	Eight 127mm (5in) guns
Aircraft:	30
Complement:	785
Launched:	1936

TIMELINE

1931   1936   1938

Graf Zeppelin

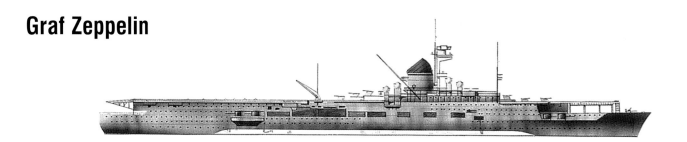

Construction of *Graf Zeppelin* to Wilhelm Hadeler's design began in 1935, but work was greatly delayed to make way for the U-boat programme. Never completed, the carrier was scuttled a few months before the end of World War II. It was raised by the Russians, but sank while under tow to Leningrad.

SPECIFICATIONS	
Type:	German aircraft carrier
Displacement:	28,540 tonnes (28,109 tons)
Dimensions:	262.5m x 31.5m x 8.5m (861ft 3in x 103ft 4in x 27ft 10in)
Machinery:	Quadruple screws, turbines
Design speed:	35 knots
Main armament:	Twelve 104mm (4.1in) guns, sixteen 150mm (5.9in) guns
Armour:	88mm (3.5in) belt, 37mm (1.5in) hangar deck
Aircraft:	42
Launched:	December 1938

Zuikaku

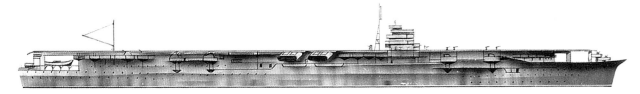

Zuikaku and its sister *Shokaku* were better armed and protected and carried more aircraft than Japan's previous purpose-built carriers. The flight deck, 240m (787ft) long and 29m (95ft) wide, was serviced by three lifts. *Zuikaku* helped sink USS *Lexington* but was sunk during the Battle of Leyte Gulf on 25 October 1944.

SPECIFICATIONS	
Type:	Japanese aircraft carrier
Displacement:	32,618 tonnes (32,105 tons)
Dimensions:	257m x 29m x 8.8m (843ft 2in x 95ft x 29ft)
Machinery:	Quadruple screws, turbines
Top speed:	34 knots
Main armament:	Sixteen 127mm (5in) guns
Armour:	175–45mm (6.5–1.8in) belt, 155–100mm (5.9–3.9in) deck
Aircraft:	84
Complement:	1660
Launched:	November 1939

Junyo

Built on a hull laid down for a passenger liner, *Junyo* carried two hangar decks, though with limited headroom. It was in action at the Aleutian Islands, Santa Cruz and the Battle of the Philippine Sea. Badly damaged by torpedoes in December 1944, it was not repaired, and was broken up in 1947.

SPECIFICATIONS	
Type:	Japanese aircraft carrier
Displacement:	24,181 tonnes (23,800 tons) full load
Dimensions:	166.55m x 21.9m x 8.05m (546ft 6in x 71ft 10in x 26ft 6in)
Machinery:	Twin screws, geared turbines
Top speed:	25 knots
Main armament:	Twelve 127mm (5in) guns
Armour:	25mm (1in) deck over machinery
Aircraft:	53
Complement:	1200
Launched:	1941

1939   1941

Japanese Aircraft & Seaplane Carriers

Seaplanes were vital for Japanese strategy, to supply garrisoned islands with no airstrip. The Japanese fleet had nine seaplane carriers, but events later in the war forced a change of role, to conventional carriers or midget submarine mother ships.

Chitose

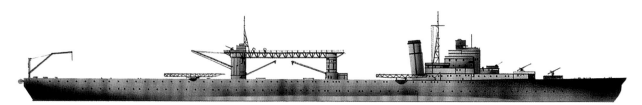

Chitose and sister-ship *Chiyoda* were seaplane carriers built for easy conversion into flush-decked aircraft carriers. In 1941 both were altered to enable the launch of midget submarines. They were refitted as conventional carriers between 1942 and 1944. *Chitose* was sunk in 1944 by aircraft from USS *Essex* and *Lexington*.

SPECIFICATIONS	
Type:	Japanese seaplane carrier
Displacement:	13,716 tonnes (13,500 tons)
Dimensions:	193m x 19m x 7m (631ft 7in x 61ft 8in x 23ft 8in)
Machinery:	Twin screws, turbines and diesel engines
Top speed:	29 knots
Main armament:	Four 127mm (5in) guns
Aircraft:	30
Launched:	November 1936

Mizuho

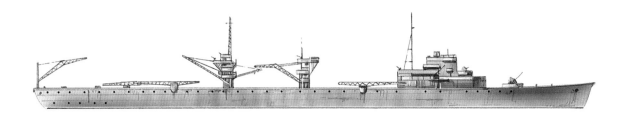

Mizuho was a sister-ship to *Chitose*, but slower, with diesel engines only. In 1941, the stern crane was removed and the ship was adapted to carry and service 12 midget submarines. There were plans for further conversion as a light aircraft carrier, but it was sunk by the US submarine *Drum* on 2 May 1942.

SPECIFICATIONS	
Type:	Seaplane (later submarine) support ship
Displacement:	11,110 tonnes (10,930 tons)
Dimensions:	192.m x 18.8m x 7.1m (631ft 6in x 61ft 8in x 23ft)
Machinery:	Twin screws, diesels
Top speed:	22 knots
Main armament:	Six 127mm (5in) guns
Aircraft:	24
Launched:	1938

TIMELINE

1936   1938   1943

Taiho

Taiho was Japan's largest purpose-built aircraft carrier. The two-tier hangars were 150m (500ft) long, unarmoured at the sides. The flight deck could withstand a 455kg (1000lb) bomb hit. Armour protection came to 8940 tonnes (8800 tons) in total. *Taiho* was sunk by the US submarine *Albacore* in June 1944.

SPECIFICATIONS	
Type:	Japanese aircraft carrier
Displacement:	37,866 tonnes (37,270 tons)
Dimensions:	260.6m x 30m x 9.6m (855ft x 98ft 6in x 31ft 6in)
Machinery:	Quadruple screws, turbines
Top speed:	33.3 knots
Main armament:	Twelve 100mm (3.9in) guns, seventy-one 25mm (1in) guns
Armour:	76mm (3in) flight deck, 150mm (5.9in) machinery
Aircraft:	53
Complement:	1751
Launched:	April 1943

Unryu

A strike carrier, for use against convoys, *Unryu* was one of only three to be completed out of a planned class of 17. With the island sponsored out, it had a wide flight deck, served by two lifts. *Unryu* was sunk by the US submarine *Redfish* in December 1944, before it saw any action.

SPECIFICATIONS	
Type:	Japanese aircraft carrier
Displacement:	22,860 tonnes (22,500 tons) full load
Dimensions:	227.4m x 27m x 7.85m (746ft 1in x 88ft 6in x 25ft 9in)
Machinery:	Quadruple screws, geared turbines
Top speed:	34 knots
Main armament:	Twelve 127mm (5in) guns
Armour:	150–45mm (5.9-1.8in) belt, 50–25mm (2–1in) deck
Aircraft:	57 + 8
Complement:	1595
Launched:	1943

Shinano

On completion, *Shinano* was the world's largest aircraft carrier, a *Yamato*-class battleship hull converted into an auxiliary carrier with vast capacity for aircraft, fuel and spare parts. It never saw active service. On 29 November 1944, it was sunk by the US submarine *Archerfish*, on the way to Kure yard for final fitting-out.

SPECIFICATIONS	
Type:	Japanese aircraft carrier
Displacement:	74,208 tonnes (73,040 tons)
Dimensions:	266m x 40m x 10.3m (872ft 9in x 131ft 3in x 33ft 9in)
Machinery:	Quadruple screws, turbines
Top speed:	27 knots
Main armament:	145 25mm (1in) guns, 16 127mm (5in) guns, 336 rocket launchers
Armour:	205mm (8.1in) belt, 190mm (7.5in) hangar deck, 80mm (3.1in) flight deck
Aircraft:	70
Complement:	2400
Launched:	October 1944

1944

Attack Cargo Ships & Commerce Raiders

As in World War I, both Britain and Germany fitted out armed merchant ships, the Germans as commerce raiders, the British as defence vessels. These could mount enough gun-power to take on a cruiser. But the lack of armour was a disadvantage.

Jervis Bay

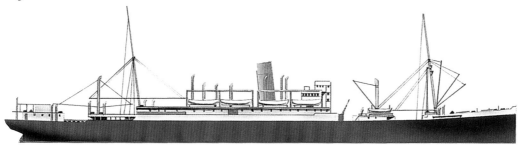

Built for the Australian emigrant trade in 1922, *Jervis Bay* was fitted out as an armed merchant cruiser with eight 152mm (6in) guns. In November 1939, its convoy was intercepted by the German battleship *Admiral Scheer*. *Jervis Bay* attacked and was sunk with heavy loss of life but the convoy was able to scatter.

SPECIFICATIONS	
Type:	British armed merchant cruiser
Displacement:	23,601 tonnes (23,230 tons)
Dimensions:	167m x 20m x 10m (549ft x 68ft x 33ft)
Machinery:	Twin screws, turbines
Top speed:	15 knots
Cargo:	Manufactured goods
Route:	London–Australian ports
Launched:	1922 (converted 1939)

Pinguin

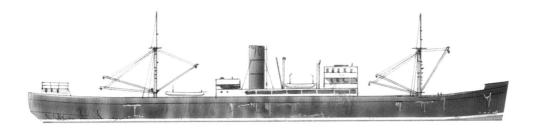

Pinguin, originally the merchant ship *Kandelfels*, was Germany's most succesful commerce raider, sinking or capturing 32 Allied vessels, totalling (145,619 tons). Carrying 420 mines and two aircraft (later one), it travelled across the world. On 8 May 1941, it was sunk off the Seychelles by the British cruiser *Cornwall*.

SPECIFICATIONS	
Type:	German commerce raider
Displacement:	17,881.6 tonnes (17,600 tons)
Dimensions:	155m x 18.7m x 8.7m (508ft 6in x 61ft 4in x 28ft 6in)
Machinery:	Twin screws, double-acting diesels
Top speed:	16 knots
Main armament:	Six 150mm (5.9in) guns, one 76mm (3in) gun
Complement:	401
Launched:	1936 (converted 1940)

TIMELINE

1922　　　　　1936　　　　　1939

Komet

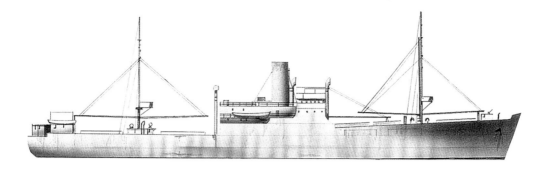

Komet was the best equipped of the German armed merchant cruisers. Armed with 150mm (5.9in) guns originally belonging to *Deutschland*-class battleships, it carried two scout planes and a motor torpedo boat (*LS2*). On its first outing, it sank 10 merchant ships, but it was sunk on a second sortie in 1942.

SPECIFICATIONS	
Type:	German commerce raider
Displacement:	7620 tonnes (7500 tons)
Dimensions:	115m x 15.3m x 6.5m (377ft 4in x 50ft 2in x 21ft 4in)
Machinery:	Single screw, diesel engines
Top speed:	14.5 knots
Main armament:	Six 150mm (5.9in) guns
Complement:	269
Launched:	1939

Kormoran

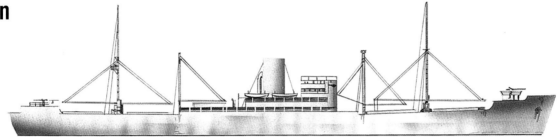

Kormoran, formerly the *Steiermark*, sank or captured 11 merchant ships totalling 69,366 tonnes (68,274 tons). Spotted by the Australian cruiser *Sydney* off Western Australia on 11 November 1941, it was taken to be a Dutch merchant vessel, until it opened fire. Both ships were destroyed in the battle.

SPECIFICATIONS	
Type:	German commerce raider
Displacement:	20,218 tonnes (19,900 tons)
Dimensions:	164m x 20m x 8.5m (538ft x 66ft 3in x 27ft 10in)
Machinery:	Twin screws, diesel engines, electric motors
Service speed:	18 knots
Main armament:	Six 150mm (5.9in) guns
Complement:	400
Launched:	1939

Artemis

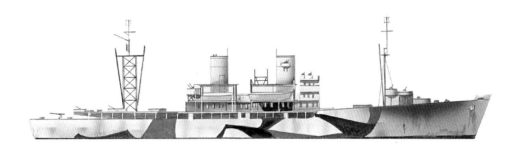

Class leader of 31 similar ships, *Artemis* was designed for invasion support, transporting 850 troops and their landing craft, which were swung down to the water, then loaded by using the ship's derricks. The hull was of wartime standard S4 pattern, able to be assembled cheaply and quickly in many yards.

SPECIFICATIONS	
Type:	US attack cargo ship
Displacement:	6848 tonnes (6740 tons) full load
Dimensions:	129.85m x 17.7m x 4.7m (426ft x 58ft x 15ft 6in)
Machinery:	Twin screws, geared turbines
Top speed:	18 knots
Main armament:	Twelve 20mm (0.78in) guns
Complement:	303
Launched:	1942

1942

Battlecruisers & Heavy Cruisers

Only the British Navy used the term 'battlecruiser' as an official designation, but other navies had similar ships. However, new forms of combat, especially aerial attack, showed this type of ship to be ill-equipped for modern warfare.

Dunkerque

Modelled on Britain's *Nelson* class to lead a fast battle-force, *Dunkerque* was a counterweight to the German *Deutschlands* of the 1930s. It carried four scout seaplanes. Damaged by British attacks at Mers-el-Kebir in July 1942, it was sailed to Toulon but scuttled in November when the Germans occupied the port.

SPECIFICATIONS	
Type:	French battlecruiser
Displacement:	36,068 tonnes (35,500 tons)
Dimensions:	214.5m x 31m x 8.6m (703ft 9in x 102ft 3in x 28ft 6in)
Machinery:	Quadruple screws, turbines
Top speed:	29.5 knots
Main armament:	Eight 330mm (13in) guns, sixteen 127mm (5in) guns
Armour:	225–125mm (8.8–4.9in) on belt, turrets 345–330mm (13.5–13in), deck 140–130mm (5.5–5in)
Complement:	1431
Launched:	October 1935

Gneisenau

Gneisenau and its companion *Scharnhorst* had their bows lengthened in 1939, improving seaworthiness. In World War II, they attacked British commerce in the North Atlantic and sank the British aircraft carrier *Glorious*. Rendered unusable by RAF bombs in 1942, *Gneisenau* was broken up between 1947 and 1951.

SPECIFICATIONS	
Type:	German battlecruiser
Displacement:	39,522 tonnes (38,900 tons)
Dimensions:	226m x 30m x 9m (741ft 6in x 98ft 5in x 30ft)
Machinery:	Triple screws, turbines
Service speed:	32 knots
Main armament:	Nine 280mm (11in) guns, twelve 150mm (5.9in) guns, fourteen 104mm (4.1 in) guns
Armour:	350–170mm (13.75–6.75in) belt, 50mm (2in) deck, 355mm (14in) main turret faces
Complement:	1840
Launched:	December 1936

TIMELINE

1935   1936   1938

Prinz Eugen

SPECIFICATIONS	
Type:	German heavy cruiser
Displacement:	19,050 tonnes (18,750 tons) full load
Dimensions:	207.7m x 21.5m x 7.2m (679ft 1in x 70ft 6in x 23ft 7in)
Machinery:	Triple screws, geared turbines
Top speed:	32.5 knots
Main armament:	Eight 203mm (8in) guns, twenty-one 105mm (4.1in) guns
Armour:	80mm (3.3in) belt, 105mm (4.1in) turret faces, 50–30mm (2–1.2in) deck
Complement:	1600
Launched:	1938

Twelve 533mm (21in) torpedo tubes and 3 aircraft added to the capabilities of this *Hipper*-class ship, commissioned in 1940. It accompanied *Bismarck,* then *Scharnhorst* and *Gneisenau,* on their Atlantic sorties. Taken over by the US Navy after the war, it was used as a target in the 1946 Bikini Atoll atom bomb tests.

Baltimore

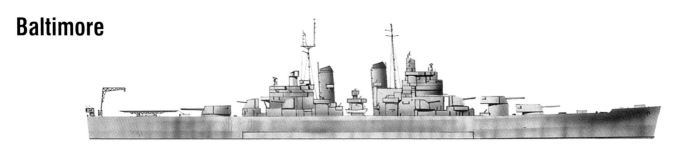

SPECIFICATIONS	
Type:	US heavy cruiser
Displacement:	17,303 tonnes (17,030 tons)
Dimensions:	205.7m x 21.5m x 7.3m (675ft 6in x 70ft 6in x 24ft)
Machinery:	Quadruple screws, geared turbines
Top speed:	33 knots
Main armament:	Nine 203mm (8in) guns
Armour:	152–102mm (6–4in) belt, 63mm (2.5in) deck
Complement:	2039
Launched:	July 1942

The *Baltimore* class were the first US cruisers built after the limits imposed by the Naval Treaties were lifted. Increased size improved sea-keeping qualities and protection. Two later became the first US guided-missile cruisers, others acted as fire support vessels in the Vietnam War. *Baltimore* was decommissioned in 1971.

Guam

SPECIFICATIONS	
Type:	US battlecruiser
Displacement:	34,801 tonnes (34,253 tons)
Dimensions:	246m x 27.6m x 9.6m (807ft 5in x 90ft 9in x 31ft 9in)
Machinery:	Quadruple screws, geared turbines
Top speed:	33 knots
Main armament:	Nine 305mm (12in) guns, twelve 127mm (5in) guns, plus 90 40mm/20mm guns
Armour:	229–127mm (9–5in) belt, 102mm (4in) deck, 330mm (13in) barbettes and turret faces
Complement:	1517
Launched:	November 1943

Larger versions of the cruiser *Baltimore*, with 305mm (12in) guns and upgraded armour, *Guam* and its sister *Alaska* were built to combat the (illusory) threat of Japanese fast raiders. *Guam* carried cranes and catapults for scout planes. Range at 15 knots was 22,800km (12,000 miles). *Guam* was scrapped in 1961.

1942

1943

Scharnhorst

A fast commerce raider, *Scharnhorst* supported the invasion of Norway in April 1940, and sank the British carrier *Glorious* in May. In February 1942, it made a famous 'dash' through the English Channel from Brest to Germany. *Scharnhorst* was sunk in the North Atlantic in December 1943 by British ships led by HMS *Duke of York*.

Scharnhorst

Scharnhorst was attacked for the next two years after the invasion of Norway by surface ships and aircraft who considered it to be a deadly threat. She remained operational despite the attacks but came to an end when attacked by *Duke of York* and three cruisers when on the way to attack an Arctic convoy.

AIRCRAFT
Four Arado Ar 196A-3 folding-wing seaplanes were carried for reconnaissance and anti-submarine defence. From 1940 one of the two catapults was removed.

SPECIFICATIONS

Type:	German battlecruiser
Displacement:	38,277 tonnes (38,900 tons)
Dimensions:	229.8m x 30m x 9.91m (753ft 11in x 98ft 5in x 32ft 6in)
Machinery:	Triple screws, geared turbines
Top speed:	32 knots
Main armament:	Nine 280mm (11in) guns, twelve 150mm (5.9in) guns
Armour:	350mm (13.8in) belt, 95mm (2.9in) deck
Crew:	1840
Launched:	30 June 1936

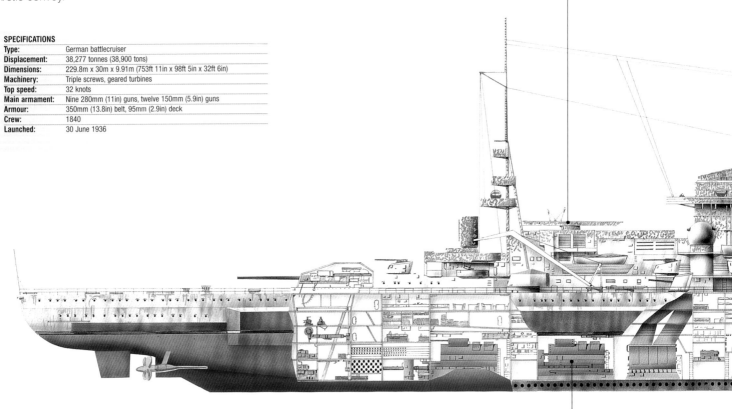

MACHINERY
Twelve Wagner 3-drum boilers raised steam for three Brown-Boveri geared turbines, with a maximum power output of 120.18MW (161,163 shp).

CONTROL TOWER
From 1940 80cm wavelength radar was fitted. This was a relatively primitive system, and radar deficiency was always a handicap to German battleships.

GUNS
A proposal to replace the main guns with 380mm (15in) guns was made but never put into action. Such armament would have produced a formidable battleship.

ARMOUR
The 357mm (14in) waterline armour belt. Like Bismarck, Scharnhorst withstood intensive torpedo and shell-fire for many hours before finally succumbing.

RANGE
Bunker capacity was 6,200 tonnes (6,101 tons) of fuel oil, giving an operational range of 18,710km (10,100 nautical miles) at 19 knots.

HULL
The peaked 'Atlantic' bow was added in 1939, but Scharnhorst remained a 'wet' ship, and heavy seas could put its 'A' turret out of action.

Battleships: Part 1

Battleships remained the prime heavy weapon of World War II fleets, with a psychological value as important as their strategic usefulness. The improvement of naval aircraft, particularly the dive bomber and the torpedo bomber, made capital ships vulnerable, and effective anti-aircraft armament became a necessity.

Admiral Graf Spee

In the 1930s, Germany built the three powerful *Deutschland* 'pocket battleships', as commerce-raiders. Savings were made by using electric welding and light alloys in the hulls. *Admiral Graf Spee* was scuttled off Montevideo, hemmed in by British ships, after the Battle of the River Plate in 1939.

SPECIFICATIONS	
Type:	German pocket battleship
Displacement:	16,218 tonnes (15,963 tons)
Dimensions:	186m x 20.6m x 7.2m (610ft 3in x 67ft 7in x 23ft 7in)
Machinery:	Twin screws, diesels
Top speed:	28 knots
Main armament:	Six 280mm (11in) guns, eight 150mm (5.9in) guns
Armour:	76mm (3in) belt, 140-76mm (5.5-3in) turrets, 38mm (1.5in) deck
Complement:	926
Launched:	April 1933

Littorio

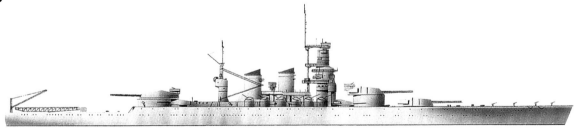

Littorio was one of the last battleships to be built for the Italian Navy. Its imposing profile was all the more striking due to the raised height of the aft turret, which was designed to avoid blast damage to the two fighter planes carried on the poop deck. *Littorio* was broken up between 1948 and 1950.

SPECIFICATIONS	
Type:	Italian battleship
Displacement:	46,698 tonnes (45,963 tons)
Dimensions:	237.8m x 32.9m x 9.6m (780ft 2in x 108ft x 31ft 6in)
Machinery:	Quadruple screws, turbines
Top speed:	28 knots
Main armament:	Nine 380mm (15in) guns, twelve 152mm (6in) guns, four 120mm (4.7in) guns, twelve 89mm (3.5in) guns
Launched:	August 1937

TIMELINE   1933   1937

Vittorio Veneto

SPECIFICATIONS	
Type:	Italian battleship
Displacement:	46,484 tonnes (45,752 tons)
Dimensions:	237.8m x 32.9m x 9.6m (780ft 2in x 108ft x 31ft 6in)
Machinery:	Quadruple screws, turbines
Top speed:	31.4 knots
Main armament:	Nine 381mm (15in) guns, twelve 152mm (6in) guns, four 120mm (4.7in) guns, twelve 89mm (3.5in) guns
Armour:	280mm (11in) belt, 162–45mm (6.4–1.8in) decks, 350–280mm (13.8–11in) barbettes, 350mm (13.8in) turret faces
Complement:	1830
Launched:	July 1937

Littorio, *Roma* and *Vittorio Veneto* made up a formidable trio of battleships. Torpedoed at the Battle of Matapan in March 1941 and again in December, *Vittorio Veneto* was damaged but repaired. Following the Italian surrender, it was laid up in the Suez Canal. It was broken up in Britain between 1948 and 1950.

North Carolina

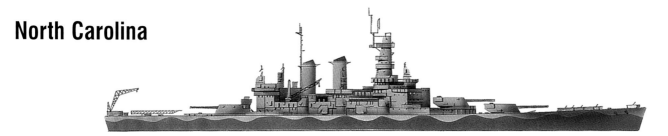

SPECIFICATIONS	
Type:	US battleship
Displacement:	47,518 tonnes (46,770 tons)
Dimensions:	222m x 33m x 10m (728ft 9in x 108ft 3in x 32ft 10in)
Machinery:	Quadruple screws, geared turbines
Top speed:	28 knots
Main armament:	Nine 406mm (16in) guns, twenty 127mm (5in) guns
Armour:	305mm (12in) belt, 140mm (5.5in) deck, 406–373mm (16–14.7in) barbettes, 406mm (16in) turret faces
Complement:	1793
Launched:	June 1940

North Carolina's design specified 355mm (14in) guns, but as the Japanese refused to restrict their main armament, it was fitted with triple 406mm (16in) gun turrets. By 1945, these had been added to by 96 x 40mm (1.6in) and 36 x 20mm (0.8in) guns. Stricken in 1960, it is preserved at Wilmington, North Carolina.

Iowa

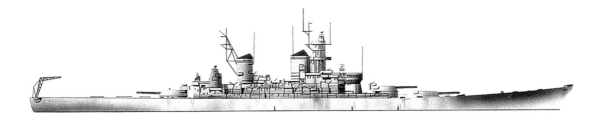

SPECIFICATIONS	
Type:	US battleship
Displacement:	56,601 tonnes (55,710 tons)
Dimensions:	270.4m x 33.5m x 11.6m (887ft 2in x 108ft 3in x 38ft)
Machinery:	Quadruple screws, turbines
Top speed:	32.5 knots
Main armament:	Nine 406mm (16in) guns, twenty 127mm (5in) guns
Armour:	310mm (12.2in) belt, 152mm (6in) deck, 440–287mm (17.3–11.3in) barbettes, 500mm (19.7in) turret faces
Complement:	1921
Launched:	August 1942

The Iowa class of fast battleships, America's last and largest, began in 1936 and had more power and heavier armour protection than the previous *South Dakota* class. A carrier escort in World War II, *Iowa* was reactivated in 1951–58 and in 1984–90 for bombardment of shore targets.

1940    1942

Bismarck

Bismarck was in most respects a modern warship. But dated armour configuration meant that the steering gear and much of the communications and control systems were poorly protected. In May 1941, on a raiding mission into the Atlantic, it sank HMS *Hood*, before being sunk itself by a British force of battleships and cruisers.

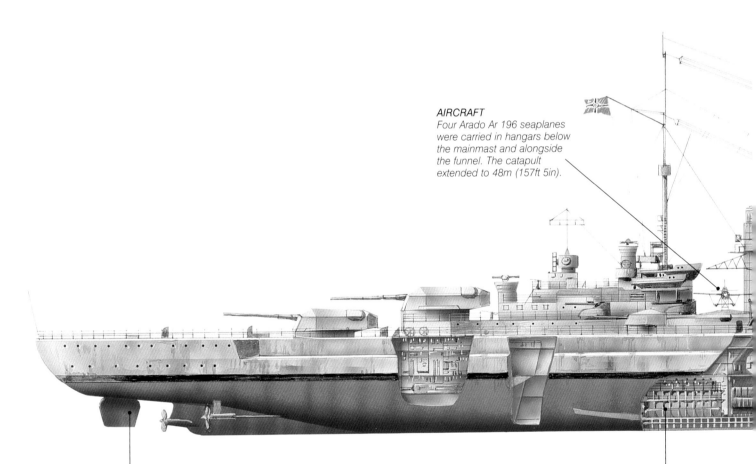

AIRCRAFT
Four Arado Ar 196 seaplanes were carried in hangars below the mainmast and alongside the funnel. The catapult extended to 48m (157ft 5in).

RUDDERS
The two parallel rudders made the ship highly manoeuvrable at speed. With an area of 24.2m² (260.5sq ft) each, they were at 80 divergence towards the centreline.

MACHINERY
Twelve Wagner Hochdruck (high pressure) oil-fired boilers powered three sets of Blohm & Voss turbines. Maximum power output was 111.92MW (150,170 shhp).

Bismarck

The 1919 Treaty of Versailles imposed tight restrictions on German naval developments. In spite of this, the Germans managed to carry out secret design studies, and when the Anglo-German Naval Treaty of 1935 came into force, were able to respond quickly. *Bismarck* and *Tirpitz* were constructed.

SPECIFICATIONS	
Type:	German battleship
Displacement:	50,955 tonnes (50,153 tons)
Dimensions:	250m x 36m x 9m (823ft 6in x118ft x 29ft 6in)
Machinery:	Three screws, geared turbines
Top speed:	29 knots
Main armament:	Eight 380mm (15in) guns, twelve 150mm (5.9in) guns
Armour:	318–267mm (12.5–10.5in) belt, 362–178mm (14.25–7in) main turrets, 121mm (4.75in) deck
Complement:	2092
Launched:	February 1939

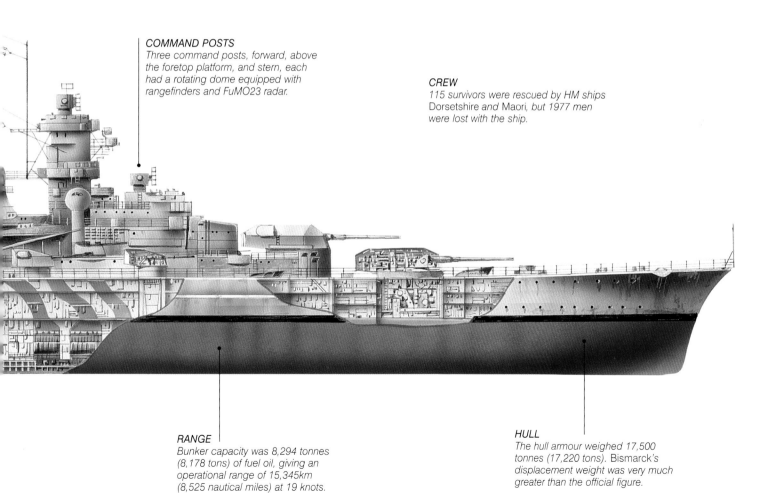

COMMAND POSTS
Three command posts, forward, above the foretop platform, and stern, each had a rotating dome equipped with rangefinders and FuMO23 radar.

CREW
115 survivors were rescued by HM ships Dorsetshire *and* Maori, *but 1977 men were lost with the ship.*

RANGE
Bunker capacity was 8,294 tonnes (8,178 tons) of fuel oil, giving an operational range of 15,345km (8,525 nautical miles) at 19 knots.

HULL
The hull armour weighed 17,500 tonnes (17,220 tons). Bismarck's displacement weight was very much greater than the official figure.

Battleships: Part 2

Although battleships took part in some of the Pacific battles, their primary role in the latter years of World War II, and in subsequent wars in Korea and Vietnam, was to bombard shore locations from long range with guns that could fire 1225kg (2700lb) high explosive shells at a range of 32km (20 miles) or more.

Washington

Washington's plans complied with the London Treaty, but when Japan refused to ratify the agreement, 406mm (16in) guns were installed, reducing top speed by 2 knots. With *South Dakota*, it sank the Japanese battlecruiser *Kirishima* at Guadalcanal in November 1942. *Washington* was scrapped in 1960/61.

SPECIFICATIONS	
Type:	US battleship
Displacement:	47,518 tonnes (46,770 tons)
Dimensions:	222m x 33m x 10m (728ft 9in x 108ft 4in x 33ft)
Machinery:	Quadruple screws, turbines
Top speed:	28 knots
Main armament:	Nine 406mm (16in) guns, twenty 127mm (5in) guns
Armour:	168–305mm (6.6–12in) belt, 178–406mm (7–16in) main turrets
Launched:	June 1940

Howe

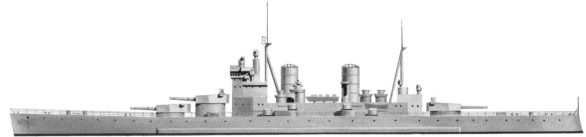

Much care was given to *Howe*'s torpedo protection. A sister-ship, *Prince of Wales,* was sunk by aerial torpedoes even before *Howe* was commissioned in 1942. It supported invasion forces in Sicily and Italy, and in 1945 was the British flagship in the Pacific. Placed in reserve in 1951, it was broken up in 1957.

SPECIFICATIONS	
Type:	British battleship
Displacement:	42,784 tonnes (42,075 tons)
Dimensions:	227.05m x 31.4m x 9.5m (745ft x 103ft x 32ft 7in)
Machinery:	Quadruple screws, geared turbines
Top speed:	28 knots
Main armament:	Ten 356mm (14in) guns, sixteen 133mm (5.25in) guns
Armour:	380–356mm (15–13in) belt, 356mm (13in) barbettes and turret faces
Complement:	1422
Launched:	1940

TIMELINE

1940 1941

Indiana

Indiana was the second of the four *South Dakota* class ships. Its 127mm (5in) guns were on two levels amidships, and the single funnel was faired into the rear of the bridge. In World War II, it was deployed mostly for carrier protection and shore bombardment. Decommissioned in 1947, it was scrapped in 1963.

SPECIFICATIONS	
Type:	US battleship
Displacement:	45,231 tonnes (44,519 tons)
Dimensions:	207.2m x 32.9m x 10.6m (680ft x 108ft x 35ft)
Machinery:	Quadruple screws, geared turbines
Top speed:	28 knots
Main armament:	Nine 406mm (16in) guns, twenty 127mm (5in) guns
Armour:	309mm (12.2in) belt, 457mm (18in) turret facings
Complement:	1793
Launched:	November 1941

Clemenceau

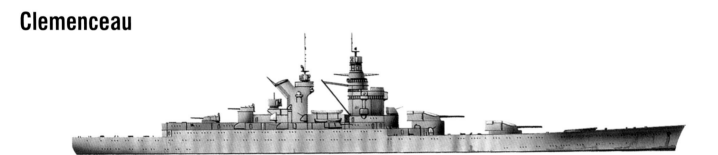

Clemenceau was only partially completed at Brest when the Germans arrived in 1940. After the D-Day invasion of 1944, they considered using it to block the harbour entrance. Before this could be done, it was sunk during a bombing raid in August. The illustration shows its appearance as indicated by the 1940 plans.

SPECIFICATIONS	
Type:	French battleship
Displacement:	48,260 tonnes (47,500 tons)
Dimensions:	247.9m x 33m x 9.6m (813ft 2in x 108ft 3in x 31ft 7in)
Machinery:	Four screws, geared turbines
Top speed:	25 knots
Main armament:	Eight 381mm (15in) guns
Launched:	1943

Vanguard

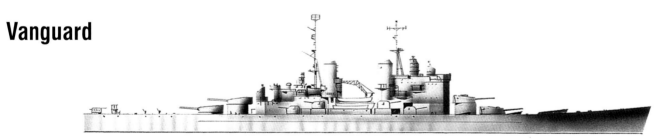

The last, largest and fastest battleship built for the Royal Navy, *Vanguard* was ordered in 1941 but did not enter service until 1946. It had a long transom stern and considerable sheer forward. Though superior to its contemporaries, the concept of the big-gun battleship was now outmoded. It was scrapped in 1960.

SPECIFICATIONS	
Type:	British battleship
Displacement:	52,243 tonnes (51,420 tons)
Dimensions:	248m x 32.9m x 10.9m (813ft 8in x 108ft x 36ft)
Machinery:	Quadruple screws, geared turbines
Top speed:	30 knots
Main armament:	Eight 380mm (15in) guns, sixteen 140mm (5.5in) guns
Armour:	114–355mm (4.5–14in) belt, 152–330mm (6–13in) on main turrets, 280–330mm (11–13in) on barbettes
Complement:	1893
Launched:	1944

1943    1944

Yamato

Yamato and *Musashi* were the largest, most powerful battleships ever. Each main turret weighed 2818 tonnes (2774 tons), and the 460mm (18.1in) guns could fire two 1473kg (3240lb) shells per minute over a distance of 41,148m (45,000yd). Used just once, they sank an escort carrier and a destroyer in 1944. *Yamato* was sunk in 1945.

AA PROTECTION
By March 1944 24 127mm and 162 25mm anti-aircraft guns were fitted. But ten aerial torpedoes and 23 bombs destroyed Yamato on 7 April 1945.

NARROW ESCAPE
During the Battle of Leyte Gulf, a narrowly missed encounter with six American battleships would have severely tested Yamato's intended ability to deal with several opponents at once.

AIRCRAFT
Yamato could carry up to seven aircraft, launched from two catapults mounted at the stern, with a retrieval crane.

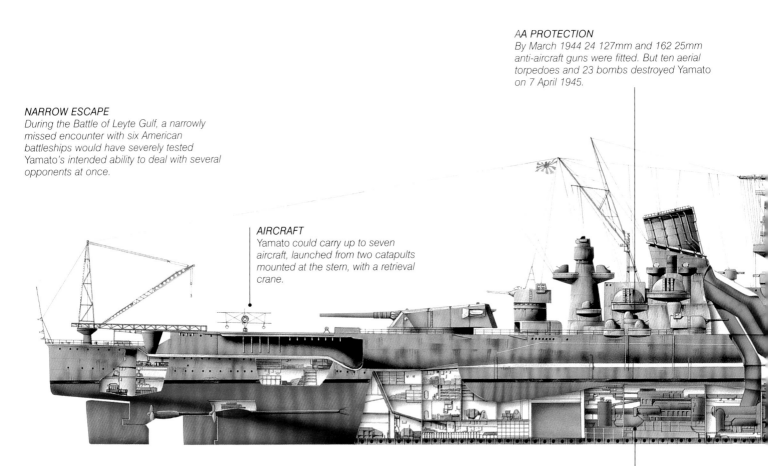

MACHINERY
At the design stage, it was planned to provide both steam turbine and diesel (or even diesel-only) power, but this was dropped.

Yamato

As flagship of the Combined Fleet *Yamato* saw action in the Battles of Midway, the Philippine Sea and Leyte Gulf. On 7th April 1945 it was sunk by US carrier aircraft 130 miles southwest of Kagoshima with the loss of 2498 lives.`

SPECIFICATIONS	
Type:	Japanese battleship
Displacement:	71,110 tonnes (71,659 tons)
Dimensions:	263m x 36.9m x 10.3m (862ft 10in x 121ft x 34ft)
Machinery:	Quadruple screws, turbines
Top speed:	27 knots
Main armament:	Nine 460mm (18.1in) guns, twelve 155mm (6.1 in) guns, twelve 127mm (5in) guns
Armour:	408mm (16.1in) belt, 546mm (21.5in) on barbettes, 193–650mm (7.6–25.6in) on main turrets, 200–231mm (7.9–9.1in) deck
Launched:	August 1940

CONTROL TOWER
It was typical of Japanese battleship design to build a high 'pagoda' main control tower which also served as a mast.

GUNS
The 460mm (18.1 in) guns, 21.13m (69ft 4in) long, were the largest ever mounted in a warship, sending 1.64 ton shells over 42km (26 miles).

MAGAZINES
The ammunition magazines were normally floodable to prevent explosion, but only if the ship's pumps were working.

Escorts & Patrol Ships: Part 1

Peacetime navies required few escort vessels, but in wartime these were needed in very large numbers for convoy protection, coastal patrols, anti-submarine and anti-aircraft work, and as support ships for the many amphibious operations carried out in the later years of World War II. Many were converted merchant ships.

Hashidate

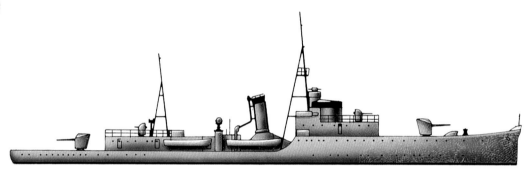

Built to support the Japanese invasion of China, this shallow-draught ship operated in coastal waters and estuaries, its main purpose to bombard shore targets. Later, *Hashidate* was refitted as an escort ship, equipped with additional light guns and AS depth-charges. It was sunk by a US submarine in May 1944.

SPECIFICATIONS

Type:	Japanese gunboat
Displacement:	1168.4 tonnes (1150 tons) full load
Dimensions:	78.5m x 9.7m x 2.45m (257ft 7in x 31ft 10in x 8ft)
Machinery:	Twin screws, geared turbines
Top speed:	19.5 knots
Main armament:	Three 120mm (4in) guns
Complement:	170
Launched:	1936

Tynwald

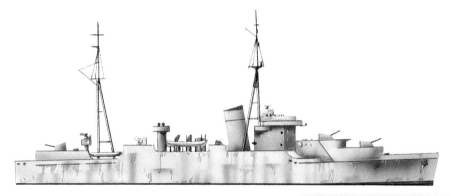

Tynwald was converted in 1940 from an Isle of Man steamer to an auxiliary AA ship, part of the Royal Navy's effort to cope with the menace of attacks from the air. Sent to the Mediterranean to support the Operation Torch landings, it was lost in November 1942, victim of a submarine or floating mine.

SPECIFICATIONS

Type:	British anti-aircraft ship
Displacement:	2474 tonnes (2376 tons) as packet ship
Dimensions:	Length 96.26m (314ft 6in)
Machinery:	Twin screw, geared turbines
Top speed:	21 knots
Main armament:	Six 102mm (4in) guns, eight 2pdr pom-pom guns
Launched:	1936

TIMELINE

1936

Erie

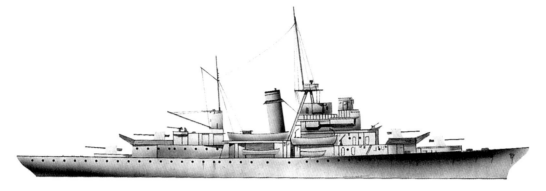

Erie and its sister *Charleston* were the first US ships to carry the new 152mm (6in), 47-calibre gun, with its combined shell and powder. The unusual hull design enabled a relatively low 5941hp (4430kW) to maintain 20 knots. A scouting plane was carried, handled by crane. *Erie* was sunk by a U-boat off Curaçao in 1942.

SPECIFICATIONS	
Type:	US gunboat
Displacement:	2376 tonnes (2339 tons)
Dimensions:	100m x 12.5m x 3.4m (328ft 6in x 41ft 3in x 11ft 4in)
Machinery:	Twin screws, turbines
Top speed:	20.4 knots
Main armament:	Four 152mm (6in) guns
Complement:	236
Launched:	1936

Hachijo

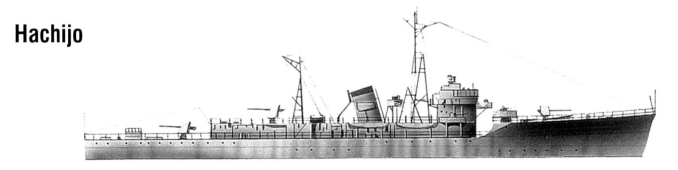

Hachijo was a prototype for successive classes of Japanese escort. During World War II, its AA armament of four 25mm (1in) guns was increased to 15; and the depth-charge load of 12 increased to 25, then 60. It was broken up in 1948.

SPECIFICATIONS	
Type:	Japanese escort
Displacement:	1020 tonnes (1004 tons)
Dimensions:	77.7m x 9m x 3m (255ft x 29ft 10in x 9ft Win)
Machinery:	Twin screws, diesel engines
Top speed:	19.7 knots
Main armament:	Three 120mm (4.7in) guns
Launched:	April 1940

Bombarda

Bombarda was one of a class of 59 escorts built to answer Italy's need for anti-submarine escort ships for supply convoys. When Italy surrendered, Germany seized *Bombarda,* among other vessels. Renamed *U-206*, it was scuttled in April 1945. Later salvaged and repaired, it remained in service until 1975.

SPECIFICATIONS	
Type:	Italian escort vessel/corvette
Displacement:	740 tonnes (728 tons)
Dimensions:	64m x 9m x 2.5m (211ft x 28ft 7in x 8ft 4in)
Machinery:	Twin screw diesel engines
Top speed:	18 knots
Main armament:	One 102mm (4in) gun
Launched:	1942

1940  1942

Escorts and Patrol Ships: Part 2

Ships of this general type were variously classed, using traditional names that did not necessarily correspond with modern functions and weapons. 'Frigate' was applied to larger ships, and 'corvette' and 'sloop' to smaller escorts. Though smaller ships were quicker to build, anti-aircraft protection required a vessel of frigate size.

Danaide

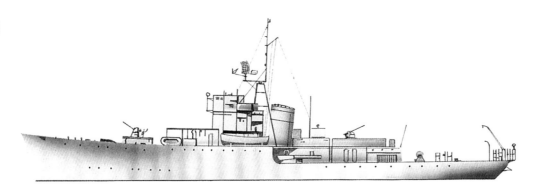

Completed only four months after launching and less than a year after being laid down, fitted for minesweeping, *Danaide* was converted to a corvette leader, with a small command deckhouse in front of the bridge replacing a depth-charge thrower. Many of its class survived the war, with 17 active in the mid-1960s.

SPECIFICATIONS	
Type:	Italian corvette
Displacement:	812 tonnes (800 tons)
Dimensions:	64m x 8.5m x 2.5m (211ft 3in x 28ft 3in x 8ft 6in)
Machinery:	Twin screws, diesel engines
Top speed:	18.5 knots
Main armament:	Four 40mm (1.6in) anti-aircraft guns
Launched:	October 1942

Avon

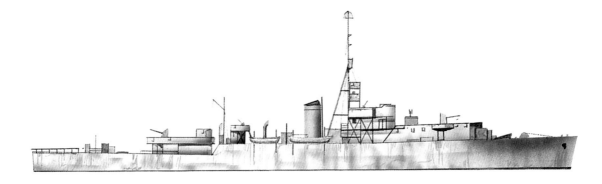

Avon was one of over 90 ocean-going, anti-submarine escorts of the 'River' class built between 1941 and 1944. Two sets of engines were installed. The original light armament was later increased. After World War II, many were passed to other navies, serving into the 1960s. *Avon* was sold to Portugal in 1949.

SPECIFICATIONS	
Type:	British frigate
Displacement:	2133 tonnes (2100 tons)
Dimensions:	91.8m x 11m x 3.8m (301ft 4in x 36ft 8in x 12ft 9in)
Machinery:	Twin screws, vertical triple expansion engines
Main armament:	Two 102mm (4in) guns
Launched:	1943

TIMELINE

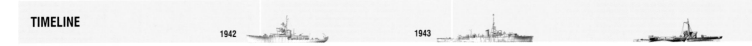

1942   1943

Daga

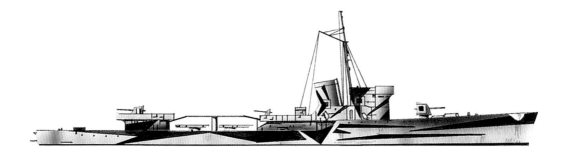

Built to protect Mediterranean convoys, 16 of these escort craft were completed before Italy left the Axis alliance, and 15, including *Daga*, were seized by the Germans in 1943. Fast but lightly armed, they were vulnerable to larger destroyers. Thirteen were sunk, including *Daga* by mines in October 1944.

SPECIFICATIONS	
Type:	Italian escort ship
Displacement:	1138 tonnes (1120 tons)
Dimensions:	82m x 8.6m x 3m (270ft x 28ft 3in x 9ft 2in)
Machinery:	Twin screws, turbines
Top speed:	31.5 knots
Main armament:	Two 100mm (3.9in) guns, six 450mm (17.7in) torpedo tubes
Launched:	July 1943

Tintagel Castle

Intended for convoys, the 'Castle' corvettes were an enlargement of the 'Flower' class but with the same machinery. *Tintagel Castle* carried the triple-barrelled Squid mortar, which used position data derived from the ship's sonar system and fired 200kg (400lb) bombs to a range of 400m (430yds). It served until 1957.

SPECIFICATIONS	
Type:	British corvette
Displacement:	1615 tonnes (1590 tons) full load
Dimensions:	76.8m x 11.2m x 4.1m (252ft x 36ft 8in x 13ft 26n)
Machinery:	Single screw, vertical triple expansion reciprocating engine
Top speed:	16.5 knots
Main armament:	One 102mm (4in) gun, one 305mm (12in) AS mortar and depth charges
Complement:	120
Launched:	July 1943

Mikura

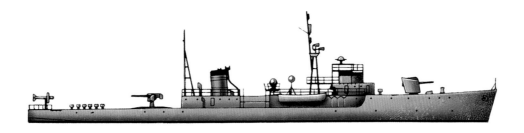

Normal construction standards were abandoned with this class of rapidly-built escort ships, whose later vessels used prefabricated parts and electric welding. But they gave good service, fitted with anti-submarine weaponry, including depth-charge racks. *Mikura*'s AA armament was much increased in 1945.

SPECIFICATIONS	
Type:	Japanese escort ship
Displacement:	1077 tonnes (1060 tons) full load
Dimensions:	78.8m x 9.1m x 3.05m (258ft 6in x 29ft 10ft)
Machinery:	Twin screws, turbines
Top speed:	19.5 knots
Main armament:	Three 120mm (4in) guns, 120 depth charges
Complement:	150
Launched:	1943

Cruisers, Part 1: American

Cruiser construction in the 1930s was, in theory, governed by the terms of the London Naval Treaty of 1930, which specified maximum armaments. The United States, which had hosted the Washington Conference of 1921–22, abided fairly scrupulously by the limitations, but built many cruisers, especially in the later 1930s.

Indianapolis

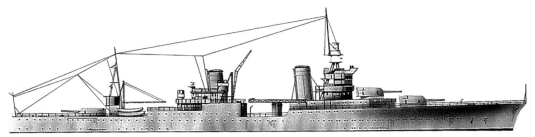

Indianapolis was the last major US surface ship to be lost in World War II. Having delivered an atom bomb (one of the two to be dropped on Japan) to the forward air base at Tinian, it was returning from the task on 29 July 1945, when it was torpedoed and sunk by a Japanese submarine.

SPECIFICATIONS	
Type:	US cruiser
Displacement:	12,960 tonnes (12,755 tons)
Dimensions:	185.9m x 20m x 6.4m (610ft x 66ft x 21ft)
Machinery:	Quadruple screws, turbines
Top speed:	32.8 knots
Main armament:	Nine 203mm (8in) guns, eight 127mm (5in) guns
Armour:	57mm (2.25in) belt, 146–63mm (5.75–2.5in) deck
Complement:	917
Launched:	November 1931

Astoria

Astoria was one of seven heavy cruisers. It saw intensive war service in 1942, escorting carrier groups and taking part in the Coral Sea and Midway battles. Part of the Northern Covering Force supporting the Guadalcanal landings, it was damaged in action against Japanese cruisers, was abandoned, and sank.

SPECIFICATIONS	
Type:	US cruiser
Displacement:	12,662 tonnes (12,463 tons)
Dimensions:	179.2m x 18.8m x 6.9m (588ft x 61ft 9in x 22ft 9in)
Machinery:	Quadruple screws, geared turbines
Top speed:	32.7 knots
Main armament:	Nine 203mm (8in) guns, eight 127mm (5in) guns
Armour:	127mm (5in) belt, 57mm (2.25in) deck
Complement:	868
Launched:	December 1933

TIMELINE

1931   1933   1936

Brooklyn

During the early 1930s, the US Navy upgraded the *Brooklyn* class cruiser design in response to the new Japanese *Mogami* class cruisers. Good protection was provided by weight saved in the hull. There were nine vessels in the new class, and all served during World War II. In 1951, *Brooklyn* was transferred to Chile.

SPECIFICATIONS	
Type:	US cruiser
Displacement:	12,395 tonnes (12,200 tons)
Dimensions:	185m x 19m x 7m (608ft 4in x 61ft 9in x 22ft 9in)
Machinery:	Quadruple screws, geared turbines
Main armament:	Fifteen 152mm (6in) guns
Launched:	November 1936

Wichita

Wichita was built on the same hull and with the same machinery as *Brooklyn* but heavier armour and guns, its 203mm (8in) guns mounted in three triple turrets. In 1945, its anti-aircraft armament was modernized. Conversion to a guided missile cruiser was considered but not effected. It was sold for breaking in 1959.

SPECIFICATIONS	
Type:	US cruiser
Displacement:	13,314 tonnes (13,015 tons)
Dimensions:	185.4m x 18.8m x 7.25m (608ft 4in x 61ft 9in x 23ft 9in)
Machinery:	Quadruple screws, geared turbines
Top speed:	33 knots
Main Armament:	Nine 203mm (8in) guns, eight 127mm (5in) guns
Armour:	152–102mm (6–4in) belt, 178mm (7in) barbettes, 203mm (8in) turret faces, 58mm (2.25in) deck
Complement:	929
Launched:	1937

Atlanta

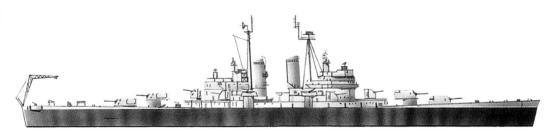

Atlanta was the lead ship in a class of 11, intended to patrol the exposed areas of the battle fleet as anti-aircraft vessels. Later members of the class carried better splinter protection, more guns and sonar capability. In 1942, *Atlanta* was torpedoed and disabled by the Japanese and finally sunk by US forces.

SPECIFICATIONS	
Type:	US cruiser
Displacement:	8473 tonnes (8340 tons)
Dimensions:	165m x 16.2m x 6.2m (541ft 6in x 53ft 2in x 20ft 6in)
Machinery:	Twin screws, turbines
Top speed:	32.5 knots
Main armament:	Sixteen 127mm (5in) guns
Launched:	1941

1937   1941

Cruisers, Part 2: American & Japanese

Japan, though a signatory to the London Naval Treaty of 1930, walked out of the 1935 follow-up conference, and went on to develop the basis of a fleet that would be second to none as a naval power in the Pacific. Japanese cruisers were comprehensively armed, carrying torpedoes, mines and aircraft.

Maya

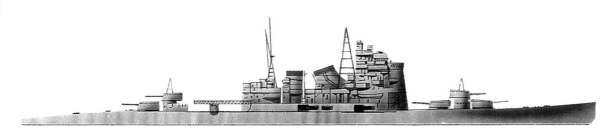

Maya had a modern look, with a high bridge. The 203mm (8in) guns had the high elevation of 70˚. In 1943, *Maya* was badly damaged at Rabaul by US aircraft, and was completely rebuilt. In October 1944, it was sunk by four torpedoes fired from a US submarine shortly before the Battle of Leyte Gulf.

SPECIFICATIONS	
Type:	Japanese cruiser
Displacement:	12,985 tonnes (12,781 tons)
Dimensions:	202m x 18m x 6m (661ft 8in x 59ft x 20ft)
Machinery:	Quadruple screws, turbines
Top speed:	35.5 knots
Main armament:	Ten 203mm (8in) guns, four 120mm (4.7in) guns
Armour:	100mm (3.9in) belt, 125mm (4.9in) magazines, 38mm (1.5in) deck
Complement:	773
Launched:	November 1930

Mogami

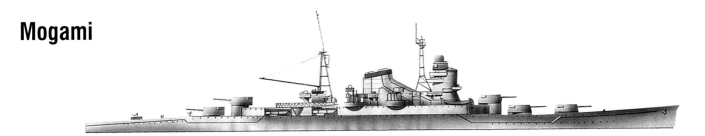

Designated a light cruiser, *Mogami* was rather more. The 155mm (6.1in) guns were positioned in triple mounts along the centreline, and the dual-purpose 127mm (5in) guns in twin mounts amidships. An initially weak hull structure was reinforced. *Mogami* was sunk in 1944 by US torpedo bombers.

SPECIFICATIONS	
Type:	Japanese cruiser
Displacement:	11,169 tonnes (10,993 tons)
Dimensions:	201.5m x 18m x 5.5m (661ft x 59ft x 18ft)
Machinery:	Quadruple screws, turbines
Top speed:	37 knots
Main armament:	Fifteen 155mm (6.1in) guns, eight 127mm (5in) dual-purpose guns
Armour:	100mm (3.9in) belt, 125mm (4.9in) magazines, 61–31.4mm (2.4–1.4in) deck
Complement:	850
Launched:	March 1934

TIMELINE

1930   1934   1937

Tone

Tone and its sister vessel Chikuma shared an unusual layout, with all main guns mounted forward, leaving the after deck clear for operating six floatplane aircraft, launched by catapult and retrieved by crane. Tone was sunk in shallow water near Kure by US aircraft in July 1945, and was broken up in situ in 1948.

SPECIFICATIONS	
Type:	Japanese cruiser
Displacement:	15,443.2 tonnes (15,200 tons)
Dimensions:	201.5m x 18.5m x 6.5m (661ft 1in x 60ft 8in x 21ft 3in)
Machinery:	Quadruple screws, geared turbines
Top speed:	35 knots
Main armament:	Eight 203mm (8in) guns, eight 127mm (5in) guns, twelve 610mm (24in) torpedo tubes
Armour:	125–100mm (4.9–3.9in) belt, 25mm (1in) turrets, 65–30mm (2.5–1.2in) deck
Complement:	850
Launched:	1937

Agano

Four ships formed the Agano class, intended as flotilla leaders for destroyer groups. Their Long Lance oxygen-powered torpedoes had three times the range of any Allied torpedo and a bigger warhead. Commissioned in October 1942, Agano was sunk by the US submarine Skate off Truk Island in 1944.

SPECIFICATIONS	
Type:	Japanese cruiser
Displacement:	8671.5 tonnes (8535 tons)
Dimensions:	174.1m x 15.2m x 5.63m (571ft 2in x 49ft 10in x 18ft 6in)
Machinery:	Quadruple screws, geared turbines
Top speed:	35 knots
Main armament:	Six 152mm (6in) guns, eight 610mm (24in) torpedo tubes
Armour:	56mm (2in) belt, 25mm (1in) turrets, 18mm (0.7in) deck
Complement:	730
Launched:	1941

Des Moines

The three ships of the Des Moines class were first to mount the complete automatic rapid-fire 203mm (8in) guns. Its two sisters were also the first warships to have air-conditioning. Des Moines remained active in the US fleet as an all-gun ship until 1961, when it was placed in reserve. It was stricken in 1991.

SPECIFICATIONS	
Type:	US cruiser
Displacement:	21,844 tonnes (21,500 tons) full load
Dimensions:	218m x 23m x 8m (717ft x 75ft 6in x 26ft)
Machinery:	Quadruple screws, turbines
Top speed:	33 knots
Main armament:	Nine 203mm (8in) guns, twelve 127mm (5in) guns
Complement:	1799
Launched:	September 1946

1941   1946

Cruisers, Part 3: British

British cruiser needs had not changed from previous years. A substantial number of middle-sized, up-to-date warships was needed to protect shipping lanes, for spells of duty at overseas stations like Simonstown (South Africa), Singapore and Hong Kong, and to maintain the balance of power against the main European navies.

Glasgow

Glasgow and four cruisers were built to match the Japanese Mogami class. Initial designs tried to reduce displacement to 8600 tonnes (8500 tons), but this was impracticable. A catapult was mounted between the funnels; a hangar for two aircraft formed an extension of the bridge. Glasgow was broken up in 1958.

SPECIFICATIONS	
Type:	British cruiser
Displacement:	11,652 tonnes (11,470 tons)
Dimensions:	187m x 19m x 5.5m (613ft 6in x 63ft x 18ft)
Machinery:	Quadruple screws, turbines
Top speed:	32 knots
Main armament:	Twelve 152mm (6in) guns
Armour:	114mm (4.5in) belt and magazines, 25mm (1in) turrets
Complement:	748
Launched:	June 1936

Belfast

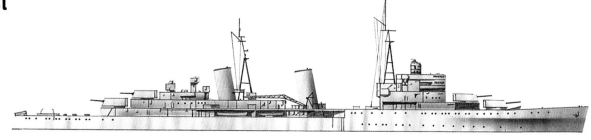

Belfast was planned as a 10,160-tonne (10,000-ton) follow-on to the Southampton class, with triple-mounted 152mm (6in) guns that could elevate to 45˚. It struck a mine four months after completion, but returned to service in October 1944, with added underwater protection. It is now a museum ship.

SPECIFICATIONS	
Type:	British cruiser
Displacement:	15,138 tonnes (14,900 tons)
Dimensions:	187m x 20m x 7m (613ft 6in x 66ft 4in x 23ft 2in)
Machinery:	Quadruple screws, geared turbines
Top speed:	32.5 knots
Main armament:	Twelve 152mm (6in) guns
Armour:	114mm (4.5in) belt, 102–51mm (3–2in) turrets, 76–51mm (3–2in) deck
Complement:	850
Launched:	1938

TIMELINE   1936   1938   1939

Dido

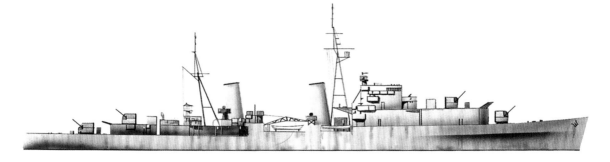

Designed for anti-aircraft defence, *Dido* carried three tiers of semi-automatic high-velocity 133mm (5.25in) guns on power-loaded mountings with a 70° elevation. A recovery crane was mounted between the funnels. Four of its class of 11 were lost during World War II, but *Dido* survived to be broken up in 1958.

SPECIFICATIONS	
Type:	British cruiser
Displacement:	6960 tonnes (6850 tons)
Dimensions:	156m x 15m x 5m (511ft 10in x 50ft 6in x 16ft 9in)
Machinery:	Quadruple screws, turbines
Top speed:	32.5 knots
Main armament:	Eight 133mm (5.25in) guns
Armour:	76mm (3in) sides, 51mm (2in) decks over magazines
Complement:	530
Launched:	July 1939

Gambia

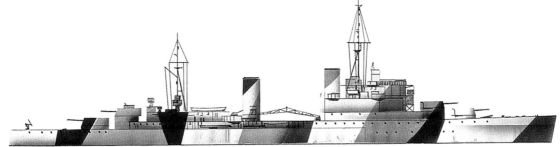

More compact than earlier British cruisers, the 'Colony' class formed the basis for later designs. Two were sunk in action in World War II, but most continued in the post-war Royal Navy, which still maintained numerous foreign stations. Two were sold to Peru in 1959, and one to India in 1957. *Gambia* was scrapped in 1968.

SPECIFICATIONS	
Type:	British cruiser
Displacement:	11,267 tonnes (11,090 tons)
Dimensions:	169.3m x 18.9m x 6.4m (555ft 6in x 62ft x 21ft)
Machinery:	Quadruple screws, turbines
Top speed:	33 knots
Main armament:	Twelve 152mm (6in) guns
Armour:	89mm (3.5in) midships belt, 51mm (2in) deck
Launched:	November 1940

Bellona

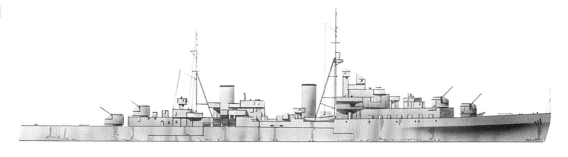

Bellona was one of five simplified *Dido* class cruisers, built as anti-aircraft ships, with new rapid-fire semi-automatic guns and lighter AA weapons. All saw active war service. After the war, *Bellona* and sister ship *Black Prince* were lent to the Royal New Zealand Navy. *Bellona* was returned in 1956 and scrapped in 1959.

SPECIFICATIONS	
Type:	British cruiser
Displacement:	7518 tonnes (7400 tons)
Dimensions:	156m x 15m x 5.4m (512ft x 50ft 6in x 18ft)
Machinery:	Quadruple screws, turbines
Top speed:	32 knots
Main armament:	Eight 133mm (5.25in) guns
Armour:	76mm (3in) side, 51–25mm (2–1in) deck
Complement:	530
Launched:	1942

1940  1942

Cruisers, Part 4: Italian

Under Mussolini's Fascist regime, Italy was aggressive and expansionist, and a powerful fleet was deemed vital. Italian cruisers were well-balanced designs, notable for their high speed, but quite lightly armed compared to their British counterparts. Their specifications strained Naval Treaty obligations to the limit – and beyond.

Fiume

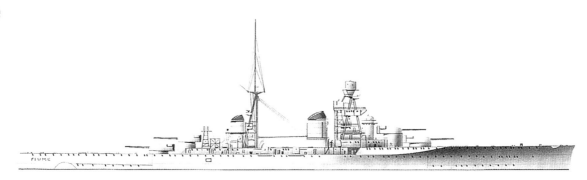

Fiume was originally designed to comply with the 10,160-tonne (10,000-ton) limit imposed by the Washington Treaty, but it was later modified. On trials the ship made 33 knots, generating 120,000hp. All four ships in the class were lost in World War II. *Fiume* was sunk in action with a British battleship in March 1941.

SPECIFICATIONS	
Type:	Italian cruiser
Displacement:	14,394 tonnes (14,168 tons)
Dimensions:	182.8m x 20.6m x 7.2m (599ft 9in x 67ft 7in x 23ft 7in)
Machinery:	Twin screws, turbines
Top speed:	32 knots
Main armament:	Eight 203mm (8in) guns
Launched:	April 1930

Giovanni delle Bande Nere

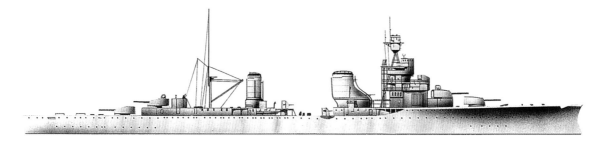

Italy built fast light cruisers to counter the large French *Jaguar*-class destroyers. Engines developed over 95,000hp, oil supply was 1250 tonnes (1230 tons) and range was 7220km (3800 miles) at 18 knots or 1843km (970 miles) at full speed. *Giovanni delle Bande Nere* was sunk by the British submarine *Urge* in April 1942.

SPECIFICATIONS	
Type:	Italian cruiser
Displacement:	6676 tonnes (6571 tons)
Dimensions:	169.3m x 15.5m x 5.3m (555ft 6in x 50ft 10in x 17ft 5in)
Machinery:	Twin screws, turbines
Main armament:	Six 100mm (3.9in), eight 152mm (6in) guns
Launched:	April 1930

TIMELINE

1930

1931

Pola

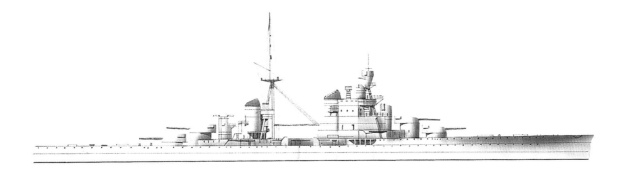

To keep *Pola*'s displacement within reasonable limits, the superstructure was reduced and the flush deck of the *Trento* class (on which the design was based) was abandoned. Unlike its sister-ships, the bridge structure and fore-funnel were combined. Additional anti-aircraft guns were added later. *Pola* was sunk in 1941.

SPECIFICATIONS	
Type:	Italian cruiser
Displacement:	13,747 tonnes (13,531 tons)
Dimensions:	182.8m x 20.6m x 7.2m (599ft 9in x 67ft 7in x 23ft 8in)
Machinery:	Twin screws, turbines
Top speed:	34.2 knots
Main armament:	Eight 203mm (8in) guns, sixteen 100mm (3.9in) guns
Launched:	December 1931

Luigi Cadorna

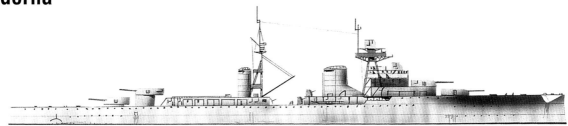

Italy's cruisers were faster than most of their contemporaries, speed achieved at the expense of protection. *Luigi Cadorna* was equipped with a new pattern of 152mm (6in) gun. It was also fitted as a minelayer, armed with 84–138 mines, according to type. *Luigi Cadorna* was removed from the effective list in 1951.

SPECIFICATIONS	
Type:	Italian cruiser
Displacement:	7113 tonnes (7001 tons)
Dimensions:	169.3m x 15.5m x 5.5m (555ft 6in x 50ft 10in x 18ft)
Machinery:	Twin screws, turbines
Main armament:	Eight 152mm (6in) guns, six 100mm (3.9in) guns
Launched:	1931

Giuseppe Garibaldi

Giuseppe Garibaldi was one of the fifth group of cruisers of the 'Condottiere' type. Between 1957 and 1962, it was rebuilt and rearmed with a twin Terrier surface-to-air missile launcher and tubes for four Polaris missiles. Although the missiles were never carried, it was the only surface vessel to be so fitted.

SPECIFICATIONS	
Type:	Italian cruiser
Displacement:	11,485 tonnes (11,305 tons)
Dimensions:	187m x 18.9m x 6.7m (613ft 6in x 62ft x 22ft)
Machinery:	Twin screws, turbines
Top speed:	30 knots
Main armament:	Four 134mm (5.3in) guns, twin Terrier missile launcher, four Polaris missile launchers (later)
Launched:	April 1934

 1934

Cruisers, Part 5: Italian & Russian

Most cruisers built in the 1930s and '40s carried at least one aircraft, launched by crane or catapult, and retrieved by crane. Crane and catapult are both apparent on the Russian *Krasnyi Kavkaz.* The point was to extend the ship's observation range, though cruiser-borne aircraft could also carry heavy machine-guns and a torpedo.

Krasnyi Kavkaz

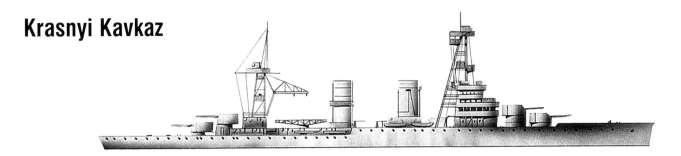

Laid down for the Russian Navy in 1913, *Krasnyi Kavkaz* was not completed until 1932, much modified from the original design. An aircraft crane formed an integral part of the mainmast. It served throughout World War II, suffering severe combat damage. It was finally sunk in 1956, as a target for the SSN-1 missile.

SPECIFICATIONS	
Type:	Soviet cruiser
Displacement:	9174 tonnes (9030 tons)
Dimensions:	169.5m x 15.7m x 6.2m (556ft x 51ft 6in x 20ft 4in)
Machinery:	Twin screws, turbines
Top speed:	29 knots
Main armament:	Four 180mm (7.1in) guns, four 100mm (3.9in) guns
Completed:	1932

Emanuele Filiberto Duca D'Aosta

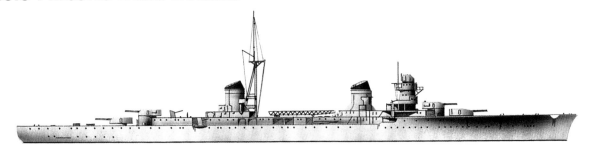

An enlargement of the *Montecuccoli* sub-class, it had the same armament but more powerful machinery and improved armour protection. *Emanuele Filiberto Duca D'Aosta* served extensively with Mediterranean convoys. Ceded to Russia in 1949, it was renamed *Stalingrad*, then *Kerch*, and discarded in the mid-1950s.

SPECIFICATIONS	
Type:	Italian cruiser
Displacement:	10,540 tonnes (10,374 tons)
Dimensions:	187m x 17.5m x 6.5m (613ft 2in x 57ft 5in x 21ft 4in)
Machinery:	Twin screws, turbines
Top speed:	37.3 knots
Main armament:	Eight 152mm (6in) guns
Armour:	70mm (2.5in) belt, 90mm (3.5in) turrets
Complement:	694
Launched:	July 1935

TIMELINE

1932   1935

Eugenio di Savoia

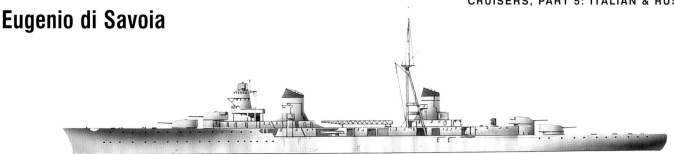

Eugenio di Savoia was one of two vessels provided for by the Italian naval programme of 1931–33. It was laid down at Ansaldo's yard in Genoa in July 1933 and completed in January 1936. In 1951, it was removed from the effective list, transferred to Greece, and renamed *Hella*. It remained in service until 1964.

SPECIFICATIONS	
Type:	Italian cruiser
Displacement:	10,842 tonnes (10,672 tons)
Dimensions:	186.9m x 17.5m x 6.5m (613ft 2in x 57ft 5in x 21ft 4in)
Machinery:	Twin screws, turbines
Top speed:	36.5 knots
Main armament:	Eight 152mm (6in) guns, six 100mm (3.9in) anti-aircraft guns, six 533mm (21in) torpedo tubes
Launched:	March 1935

Kirov

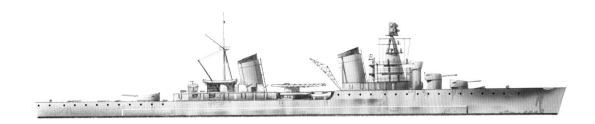

The Soviet Union's first home-built cruiser, *Kirov* was completed in 1938. The 180mm (7.1) guns were mounted in triple turrets, with the battery of 100mm (3.9in) guns in single mounts alongside the second funnel. A catapult was sited between the two funnels. *Kirov* was deleted from the list in the late 1970s.

SPECIFICATIONS	
Type:	Soviet cruiser
Displacement:	11,684 tonnes (11,500 tons)
Dimensions:	191m x 18m x 6m (626ft 8in x 59ft x 20ft)
Machinery:	Twin screws, turbines
Top speed:	35.9 knots
Main armament:	Nine 180mm (7.1in) guns, six 100mm (3.9in) guns
Armour:	50mm (2in) belt and deck, 75mm (3in) turrets
Complement:	734
Launched:	November 1936

Etna

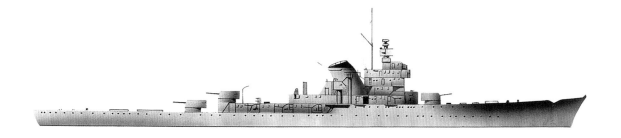

Ordered for Siam in 1938, work on *Etna* ceased in December 1941, and in 1942 it was taken over by the Italian Navy. Major changes were introduced to turn it into an anti-aircraft cruiser. It was little over half complete when seized by German forces in 1943, but was finally scuttled that year, at Trieste.

SPECIFICATIONS	
Type:	Italian cruiser
Displacement:	5994 tonnes (5900 tons)
Dimensions:	153.8m x 14.4m x 5.9m (504ft 7in x 47ft 6in x 19ft 6in)
Machinery:	Twin screws, turbines
Top speed:	28 knots
Main armament:	Six 135mm (5.3in) guns
Launched:	May 1942

1936

1942

Cruisers, Part 6: Other Navies

Cruisers in the 1930s and '40s were powered by turbine engines, in which steam from the boilers turns blades attached to a shaft. A cruiser's engines developed 60,246–67,770kW (80,000–90,000hp), a destroyer only 37,650kW (50,000hp). Gearing to the shafts made the ship economical at low speeds, extending its range.

Dupleix

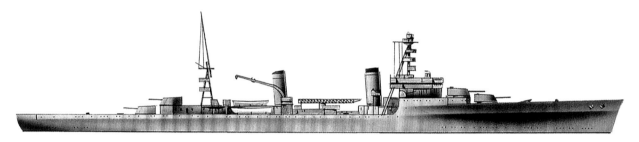

Dupleix was the last of a group of four cruisers laid down each year from 1926 to 1929, modified versions of the *Tourville* class, and about two knots were sacrificed in favour of better protection. Scuttled at Toulon in 1942, *Dupleix* was raised in 1943, only to be sunk again by Allied bombing.

SPECIFICATIONS

Type:	French cruiser
Displacement:	12,984 tonnes (12,780 tons)
Dimensions:	194m x 19.8m x 7m (636ft 6in x 65ft x 23ft 7in)
Machinery:	Triple screws, turbines
Top speed:	34 knots
Main armament:	Eight 203mm (8in) guns, eight 89mm (3.5in) guns
Armour:	51–57mm (2–2.25in) belt
Launched:	October 1930

Delhi

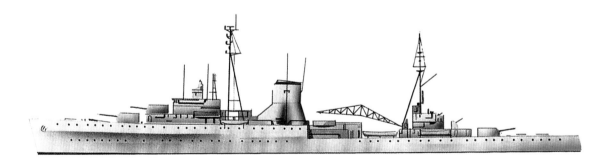

Delhi, then the British *Leander* class cruiser *Achilles*, helped to defeat the German battleship *Admiral Graf Spee* in December 1939. In 1948, it was sold to the Indian Navy, serving as the flagship until 1957. In 1959, when India began to modernize its fleet, *Delhi* was put into general service. It was scrapped in 1978.

SPECIFICATIONS

Type:	Indian cruiser
Displacement:	9895 tonnes (9740 tons)
Dimensions:	166m x 16.7m x 6m (544ft 6in x 55ft 2in x 20ft)
Machinery:	Quadruple screws, turbines
Top speed:	32 knots
Main armament:	Six 152mm (6in) guns, four 102mm (4in) anti-aircraft guns
Launched:	September 1932

TIMELINE

1930   1932

Baleares

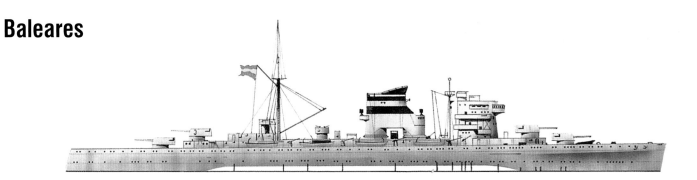

Designed by Sir Philip Watts, *Baleares* followed the British *Kent* class but with better speed and improved anti-aircraft armament. In the Spanish Civil War, it formed part of the Nationalist force and in 1938 encountered Government ships off Cape Palos, where it was torpedoed and sank with heavy loss of life.

SPECIFICATIONS	
Type:	Spanish cruiser
Displacement:	13,279 tonnes (13,070 tons)
Dimensions:	193.5m x 19.5m x 5.2m (635ft x 64ft x 17ft 4in)
Machinery:	Quadruple screws, geared turbines
Main armament:	Eight 203mm (8in) guns
Armour:	51mm (2in) belt, 114mm (4.5in) over magazines
Complement:	780
Launched:	1932

Gloire

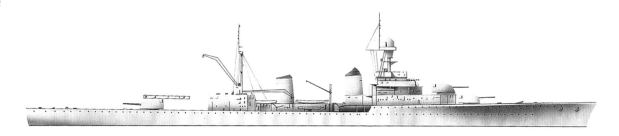

Gloire's 152mm (6in) guns were mounted in triple turrets, two superfiring forward and one aft. The aft superstructure comprised a large hangar with full repair facilities, and a long, open deck provided space to handle four aircraft. On top of the aft turret was a launching catapult. *Gloire* was scrapped in 1958.

SPECIFICATIONS	
Type:	French cruiser
Displacement:	9245 tonnes (9100 tons)
Dimensions:	179.5m x 17.4m x 5.3m (589ft x 57ft 4in x 17ft 7in)
Machinery:	Twin screws, turbines
Top speed:	36 knots
Main armament:	Nine 152mm (6in) guns, eight 88mm (3.5in) guns
Armour:	102mm (4in) belt
Launched:	1935

De Grasse

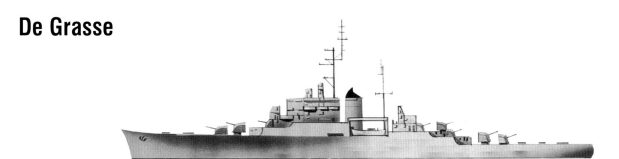

Its construction suspended during the German occupation of France. *De Grasse was* completed in 1956, as a fleet command ship for radar-controlled air strikes. In 1966, it was refitted to serve the Pacific Experimental Nuclear Centre. Several gun turrets were removed, and a lattice communications mast was fitted aft.

SPECIFICATIONS	
Type:	French cruiser
Displacement:	11,730 tonnes (11,545 tons)
Dimensions:	188m x 18.5m x 5.4m (617ft 2in x 61ft x 18ft 2in)
Machinery:	Twin screws, turbines
Top speed:	33.5 knots
Main armament:	Twelve 127mm (5in) dual-purpose guns
Launched:	1946

1935   1946

De Ruyter

Enlarged during construction but still rather lightly armed, and stationed in the Dutch East Indies, this ship saw intensive action against Japanese forces. In February 1942 it was flagship of the Allied squadron opposing the Japanese invasion of Indonesia. It was sunk by torpedo in the Battle of the Java Sea, on 27 February 1942.

ANTI-AIRCRAFT GUNS
The ten Bofors 40mm (1.5in) anti-aircraft guns, combined with a highly-effective fire-control system, were among the ship's best features.

AIRCRAFT
The floatplanes were launched from a midships-installed Heinkel K8 catapult, and retrieved from the sea by crane.

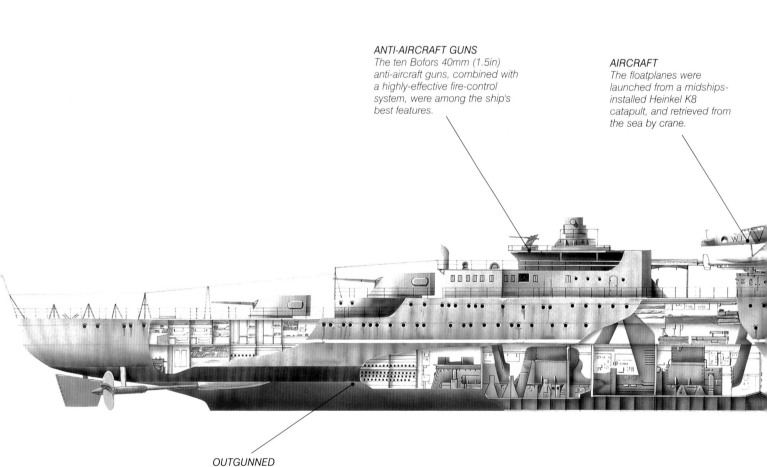

OUTGUNNED
De Ruyter and its squadron were outgunned and outnumbered in the Battle of the Java Sea. A Japanese 'long lance' torpedo was responsible for sinking it.

De Ruyter

De Ruyter and De Zeven Provincien both took part in several NATO exercises and were often used as flagships for different naval task forces. De Zeven Provincien underwent a refit between 1962 and 1964 but a lack of funds meant De Ruyter did not undergo the same changes. De Ruyter was decommissioned in 1973.

SPECIFICATIONS

Type:	Dutch cruiser
Displacement:	6,650 tonnes (6,545 tons)
Dimensions:	170.9m x 15.7m x 5.1m (561ft x52ft x 17ft)
Machinery:	Three screws, geared turbines
Top speed:	32 knots
Main armament:	Seven 150mm (6in) guns, ten 40mm (1.5in) guns
Armour:	50mm (2in) belt, 30mm (1.2in) deck and turrets
Aircraft:	Two Fokker C-11W floatplanes
Complement:	435
Launched:	1935

CONNING TOWER
The 'conning tower' design was based on German models. It provided somewhat cramped accommodation for a flagship's staff and functions.

COMMUNICATIONS
The lack of masts was an unusual feature, and a stack extension was needed to support radio antennae. The stack cap was replaced after commissioning tests.

ARMAMENT
The original design had only six main guns. A seventh was added during construction but the ship remained lightly gunned.

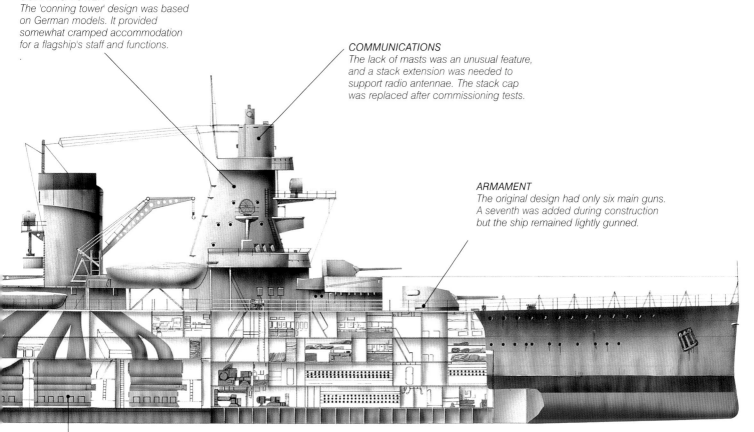

MACHINERY
Six boilers and three geared steam turbines generated 66,000shp (49,216 kW). 1,300 tonnes of oil gave a range of 6,800 nm (12,594km) at 12 knots.

Destroyers, Part 1: Russian

In the course of the 1930s, Soviet Russia set about expanding and modernizing its destroyer fleet, in successive five-year plans. Italian technical help was used and Soviet destroyers of the early 1930s closely resembled Italian types. Forty-eight fleet destroyers were built, most armed with mines as well as guns and torpedoes.

Minsk

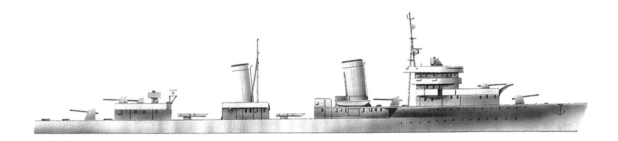

Conceived as a type of super-destroyer for carrying out raiding missions in the Baltic, *Minsk* was built with technical aid from France and Italy. It was sunk in 1941, but refloated in 1942. In 1959, it became a training vessel.

SPECIFICATIONS	
Type:	Soviet destroyer
Displacement:	2623 tonnes (2582 tons)
Dimensions:	127.5m x 11.7m x 4m (418ft 4in x 38ft 4in x 13ft 4in)
Machinery:	Triple screws, turbines
Top speed:	40 knots
Main armament:	Five 130mm (5.1in) guns, two 76mm (3in) guns
Launched:	November 1935

Bodryi

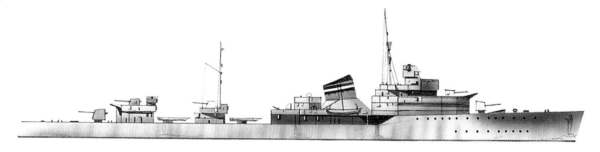

One of the first group of 28 ships, of the new Type 7, *Bodryi* was found insufficiently seaworthy for Arctic and Northern Pacific conditions. For this reason, a revised design was produced. By the start of World War II, 46 ships were complete, of which 20 were later lost. *Bodryi* was scrapped in 1958.

SPECIFICATIONS	
Type:	Soviet destroyer
Displacement:	2072 tonnes (2039 tons)
Dimensions:	113m x 10m x 4m (370ft 3in x 33ft 6in x 12ft 6in)
Machinery:	Twin screws, geared
Main armament:	Four 127mm (5.1in) guns
Launched:	1936

TIMELINE

1935   1936

Gromki

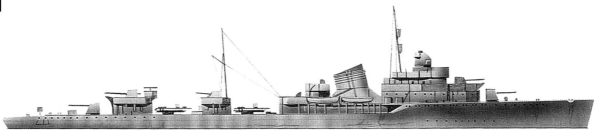

Gromki was a Type 7 destroyer, of Italian influence, completed in 1939. Its AA armament was uprated during the war. Engines developed 48,000hp, and oil fuel capacity was 548 tonnes (540 tons), enough for 1533km (807 miles) at full speed and 4955km (2608 miles) at 19 knots. Gromki was discarded in the 1950s.

SPECIFICATIONS	
Type:	Soviet destroyer
Displacement:	2070 tonnes (2039 tons)
Dimensions:	112.8m x 10.2m x 3.8m (370ft 3in x 33ft 6in x 12ft 6in)
Machinery:	Twin screws, turbines
Main armament:	Four 130mm (5.1in) guns, two 76mm (3in) guns, six 533mm (21in) torpedo tubes
Launched:	1936

Silnyi

Italian influence is evident in the design of the Type 7 destroyers. Silnyi was one of the second group, 7U, known as the Storozhevoi class, with strengthened hulls and more powerful engines. It carried 60 mines as well as the normal destroyer armament. Part of the Baltic fleet, it was withdrawn in the mid-1960s.

SPECIFICATIONS	
Type:	Soviet destroyer
Displacement:	2443.5 tonnes (2405 tons)
Dimensions:	112.8m x 10.2m x 4.1m (370ft 7in x 22ft 6in x 13ft 5in)
Machinery:	Twin screws, geared turbines
Top speed:	36 knots
Main armament:	Four 130mm (5.1in) guns, six 533mm (21in) torpedo tubes
Complement:	207
Launched:	1938

Ognevoi

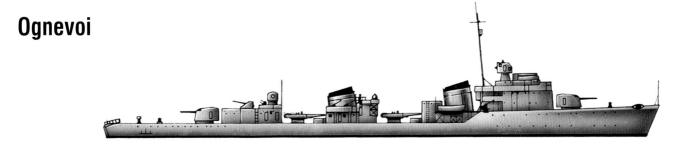

Only two ships of this two-funnelled class were completed before the end of World War II, but another 12 were built after 1945, known as the Skoryi class. Fast and well-armed, Ognevoi could carry 96 mines in addition to its secondary armament of two 76mm (3in) guns and four heavy 12.7mm (0.5in) machine guns.

SPECIFICATIONS	
Type:	Soviet destroyer
Displacement:	2997.2 tonnes (2950 tons)
Dimensions:	117m x 11m x 4.2m (383ft 10in x 36ft 1in x 13ft 9in)
Machinery:	Twin screws, geared turbines
Top speed:	37 knots
Main armament:	Four 130mm (5.1in) guns, six 533mm (21in) torpedo tubes
Complement:	250
Launched:	1940

1938   1940

Destroyers, Part 2: British

The typical British destroyer of the 1930s displaced about 2000 tonnes and could reach about 35 knots top speed. Considerably larger than its predecessors, its function was considered to be the same, though gradually it took on a great variety of roles. As in other navies, it was later given much improved AA defences.

Ardent

With the 'W' class, the Royal Navy began a new era of destroyer construction, after a lapse of eight years. Completed in 1930, *Ardent* was sunk in June 1940 by *Scharnhorst* and *Gneisenau*, while escorting the aircraft carrier *Glorious*, which was also sunk. Three other 'W' class ships were lost during the war.

SPECIFICATIONS	
Type:	British destroyer
Displacement:	2022 tonnes (1990 tons)
Dimensions:	95.1m x 9.8m x 3.7m (312ft x 32ft 3in x 12ft 3in)
Machinery:	Twin screws, geared turbines
Main armament:	Four 120mm (4.7in) guns, eight 533mm (21in) torpedo tubes
Top speed:	35 knots
Launched:	1929

Crescent

A 'C' class destroyer, *Crescent* had increased fuel capacity compared to the 'B' class, and a 76mm (3in) anti-aircraft gun, as well as the 120mm (4.7in) guns and eight 533mm (21in) torpedo tubes. *Crescent* was transferred to the Royal Canadian Navy in 1937, and was sunk in a collision on 25 June 1940.

SPECIFICATIONS	
Type:	British destroyer
Displacement:	1927 tonnes (1897 tons)
Dimensions:	97m x 10m x 3m (317ft 9in x 33ft x 8ft 6in)
Machinery:	Twin screws, turbines
Top speed:	36.4 knots
Main armament:	Four 120mm (4.7in) guns
Launched:	September 1931

TIMELINE

1929   1931   1932

Duncan

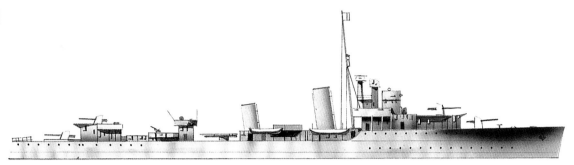

Fitted out originally as a destroyer leader, *Duncan* was a 'D' class vessel laid down in 1932, a slightly enlarged version of the 'B' class. During World War II, after the class had been greatly reduced by losses, the remaining vessels were converted into escort ships. *Duncan* was broken up in 1945.

SPECIFICATIONS	
Type:	British destroyer
Displacement:	1973 tonnes (1942 tons)
Dimensions:	100m x 10m x 4m (329ft x 32ft 10in x 12ft 10in)
Machinery:	Twin screws, turbines
Main armament:	Four 120mm (4.7in) guns
Launched:	July 1932

Exmouth

An enlarged version of the standard destroyer, with an additional 120mm (4.7in) gun and higher speed, *Exmouth* was an excellent sea boat. At 20 knots, just over 2 tonnes (2 tons) of oil fuel were consumed every hour. It was sunk with all hands in the Moray Firth in January 1940, probably by U-boat torpedo.

SPECIFICATIONS	
Type:	British destroyer leader
Displacement:	2041 tonnes (2009 tons)
Dimensions:	104.5m x 10.2m x 3.8m (342ft 10in x 33ft 9in x 12ft 6in)
Machinery:	Twin screws, turbines
Main armament:	Five 120mm (4.7in) guns
Launched:	February 1934

Cleveland

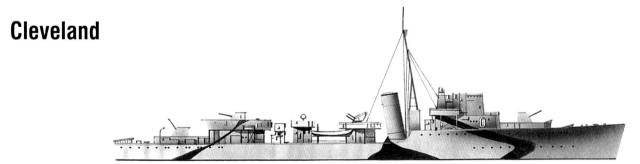

Cleveland was in the first group of the 'Hunt' class of destroyers, which eventually numbered 86. Designed to carry six 102mm (4in) guns, this armament proved too heavy, and the number was reduced to four. Though seaworthy, the ships tended to roll. *Cleveland* was wrecked en route to the breakers in June 1957.

SPECIFICATIONS	
Type:	British destroyer
Displacement:	1473 tonnes (1450 tons)
Dimensions:	85m x 9m x 4m (280ft x 29ft x 12ft 6in)
Machinery:	Twin screws, turbines
Top speed:	28 knots
Main armament:	Four 102mm (4in) guns
Launched:	April 1940

1934    1940

Cossack

A 'Tribal' class destroyer, *Cossack* made a dramatic rescue of British prisoners from the German vessel *Altmark*, in Norwegian waters, in February 1940, and was part of the force pursuing *Bismarck*, in May 1941. It also saw action with convoys to Malta, before being sunk in the North Atlantic by *U-563* on 26 October 1941.

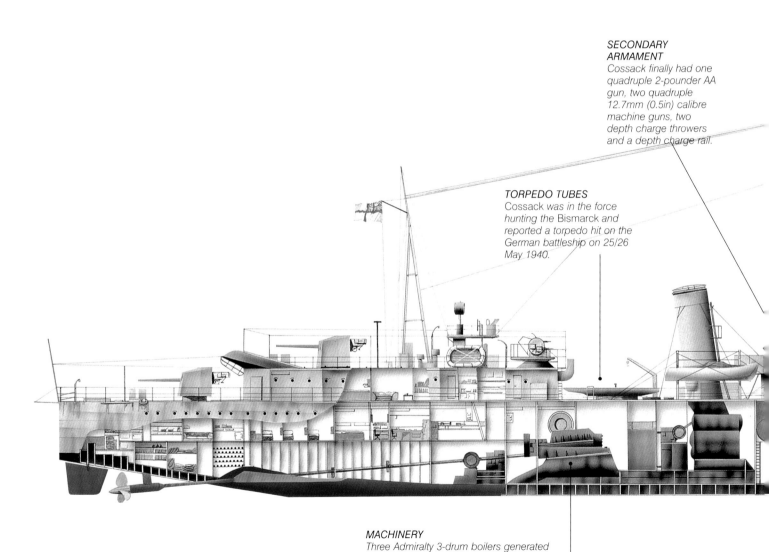

SECONDARY ARMAMENT
Cossack finally had one quadruple 2-pounder AA gun, two quadruple 12.7mm (0.5in) calibre machine guns, two depth charge throwers and a depth charge rail.

TORPEDO TUBES
Cossack was in the force hunting the Bismarck and reported a torpedo hit on the German battleship on 25/26 May 1940.

MACHINERY
Three Admiralty 3-drum boilers generated steam for two Parsons geared turbines, with a maximum power output of 33,131 kW (44,430 shp).

Cossack

In May 1941 *Cossack* participated in the pursuit and destruction of the *Bismarck*. While escorting Convoy WS-8B to the Middle East, Cossack and four other destroyers made several torpedo attacks in the evening and into the next morning. No hits were scored, but they made it easier for the battleships to attack her the next morning.

SPECIFICATIONS

Type:	British destroyer
Displacement:	1900 tonnes (1870 tons)
Dimensions:	111.5m x 11.13m x 4m (364ft 8in x 36ft 6in x 13ft)
Machinery:	Twin screws, geared turbines
Top speed:	36 knots
Main armament:	Eight 120mm (4.7in) guns, four 53mm (21in) torpedo tubes
Complement:	219
Launched:	June 1937

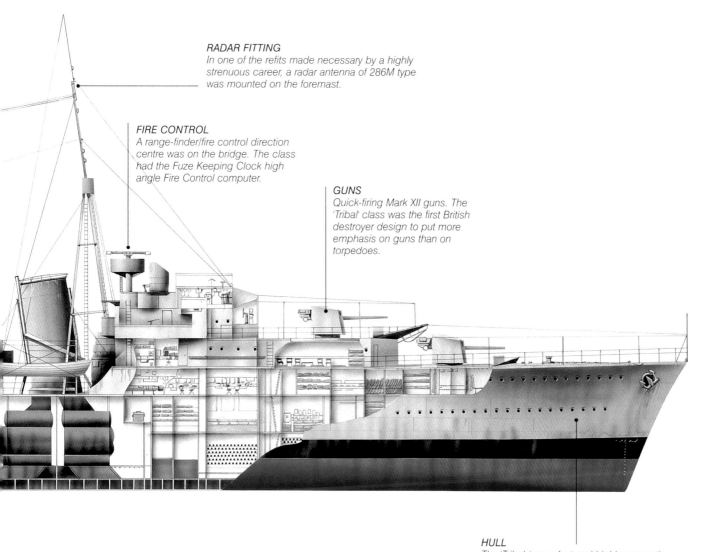

RADAR FITTING
In one of the refits made necessary by a highly strenuous career, a radar antenna of 286M type was mounted on the foremast.

FIRE CONTROL
A range-finder/fire control direction centre was on the bridge. The class had the Fuze Keeping Clock high angle Fire Control computer.

GUNS
Quick-firing Mark XII guns. The 'Tribal' class was the first British destroyer design to put more emphasis on guns than on torpedoes.

HULL
The 'Tribals' were fast and highly seaworthy ships, and looked it, with their clipper bows, and raked masts and funnels.

Destroyers, Part 3: British & French

French destroyer design was influenced by Italian design, and vice versa, the competition producing classes that verged on the dimensions of the light cruiser. British destroyers were small and lightly gunned by comparison, but perhaps more suited to the rapid tactics and tight manoeuvring vital for a destroyer's true purpose.

Aigle

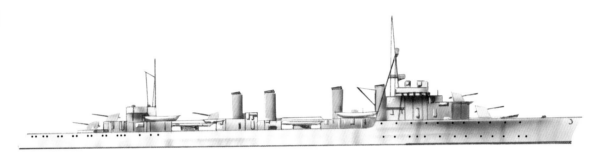

One of a class of six, the only four-funnelled destroyers of the 1930s, *Aigle* was fitted with new rapid-fire semi-automatic guns, capable of 12–15 rounds per minute. With two sister-ships, it was scuttled at Toulon in 1942 as German forces entered the port. Refloated, it was sunk again by air attack in 1943.

SPECIFICATIONS	
Type:	French destroyer
Displacement:	3190.3 tonnes (3140 tons)
Dimensions:	128.5m x 11.84m x 4.79m (421ft 7in x 38ft 10in x 16ft 4in)
Machinery:	Twin screws, geared turbines
Top speed:	36 knots
Main armament:	Five 140mm (5.5in) guns, six 550mm (21.7in) torpedo tubes
Complement:	230
Launched:	1931

L'Indomptable

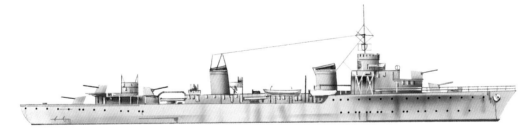

Each new Italian destroyer type prompted a bigger or better French one. The *Fantasque* class was fast, capable of over 40 knots. With *Le Triomphant* and *Le Malin*, *L'Indomptable* made a raid into the Skagerrak during the German invasion of Norway, engaging German patrol boats. It was scuttled at Toulon in 1942.

SPECIFICATIONS	
Type:	French destroyer
Displacement:	3352.8 tonnes (3300 tons)
Dimensions:	132.4m x 12.35m x 5m (434ft 4in x 40ft 6in x 16ft 4in)
Machinery:	Twin screws, geared turbines
Top speed:	40 knots
Main armament:	Five 138.6mm (5.46in) guns, three triple 533mm (21in) torpedo tubes
Complement:	210
Launched:	1933

TIMELINE

1931  1933   1941

Exmoor

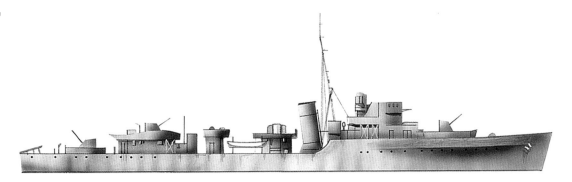

Laid down as HMS *Burton*, this ship was renamed after the first *Exmoor*, also a 'Hunt' class destroyer, was sunk by a German E-boat in February 1941. *Exmoor*, of the second, slightly larger, group of 'Hunts', was transferred to the Danish Navy in 1953, as *Valdemar Sejr*. It was broken up in 1966.

SPECIFICATIONS	
Type:	British destroyer
Displacement:	1651 tonnes (1625 tons)
Dimensions:	85.3m x 9.6m x 3.7m (280ft x 31ft 6in x 12ft 5in)
Machinery:	Twin screws, turbines
Top speed:	26.7 knots
Main armament:	Six 102mm (4in) guns
Launched:	March 1941

Comet

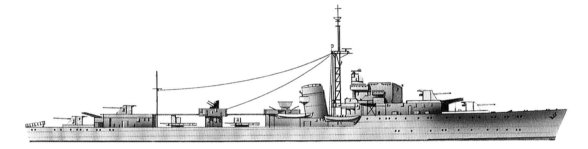

Comet belonged to a 24-unit class of large destroyers, developed late in World War II. One, HMS *Contest*, was the first British destroyer with an all-welded hull. All vessels in the class survived the war. Four were passed to Norway in 1946, and four to Pakistan in the 1950s. *Comet* was broken up in 1962.

SPECIFICATIONS	
Type:	British destroyer
Displacement:	2575 tonnes (2535 tons)
Dimensions:	111m x 11m x 4m (362ft 9in x 35ft 8in x 14ft 5in)
Machinery:	Twin screws, turbines
Top speed:	36.7 knots
Main armament:	Four 114mm (4.5in) guns
Launched:	June 1944

Daring

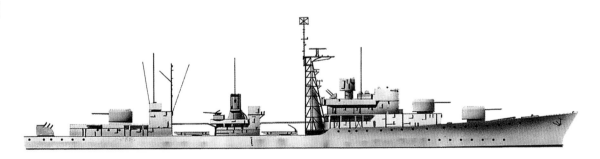

Multi-purpose ships, *Daring* and seven others were the largest destroyers yet built for the Royal Navy and had an all-welded hull construction. The lattice foremast was built around the fore funnel. The 114mm (4.5in) guns were automatic and radar controlled. *Daring* was withdrawn and broken up in 1971.

SPECIFICATIONS	
Type:	British destroyer
Displacement:	3636 tonnes (3579 tons)
Dimensions:	114m x 13m x 4m (375ft x 43ft x 13ft)
Machinery:	Twin screws, turbines
Top speed:	31.5 knots
Main armament:	Six 114mm (4.5in) guns
Launched:	August 1949

1944   1949

Destroyers, Part 4: USA

In December 1941, the United States had 68 destroyers in the Pacific and 51 in the Atlantic. In the course of 1942–45, another 302 were built, and the destroyer became the US Navy's all-purpose work-horse, as fleet and carrier escort, submarine hunter, anti-aircraft battery ship and landing support ship.

Sims

Fleet destroyers had to be fast and this required powerful engines. The *Sims* class destroyers were heavier than had been planned but were an effective class, capable of long range service, up to 12,000km (6500nm) at 12 knots. *Sims* was sunk by Japanese aircraft in the Battle of the Coral Sea, on 7 May 1942.

SPECIFICATIONS	
Type:	US destroyer
Displacement:	2388.6 tonnes (2315 tons)
Dimensions:	106.15m x 10.95m x 3.9m (348ft 4in x 36ft x 12ft 10in)
Machinery:	Twin screws, geared turbines
Top speed:	35 knots
Main armament:	Five 127mm (5in) guns, eight 533mm (21in) torpedo tubes
Complement:	192
Launched:	1938

Doyle

The large class to which *Doyle* belonged were the last destroyers designed for the United States before it entered World War II. Production was speeded up by simplified design, many being completed with straight-fronted bridge structures. Many were transferred to other navies after the war. *Doyle* was scrapped in 1970.

SPECIFICATIONS	
Type:	US destroyer
Displacement:	2621 tonnes (2580 tons)
Dimensions:	106m x 11m x 5.4m (348ft 6in x 36ft x 18ft)
Machinery:	Twin screws, turbines
Top speed:	37.4 knots
Main armament:	Four 127mm (5in) guns, five 533mm (21in) torpedo tubes
Launched:	March 1942

TIMELINE

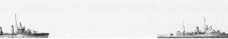

1938   1942

Edsall

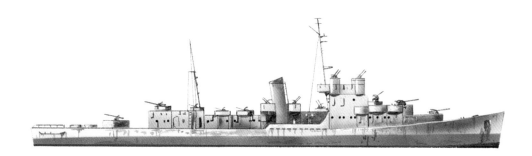

Speed was not vital for a merchant convoy escort ship, as the convoys were relatively slow; it had to be manoeuvrable and quick to respond to attack from surface craft and submarine alike. Hundreds of these ships were built in six very similar classes. *Edsall*'s diesel engines were driven through a reduction gearbox.

SPECIFICATIONS	
Type:	US destroyer escort
Displacement:	1625.6 tonnes (1600 tons) full load
Dimensions:	93.3m x 11.15m x 3.2mm (306ft x 36ft 7in x 10ft 5in)
Machinery:	Twin screws, diesels
Top speed:	21 knots
Main armament:	Three 76mm (3in) guns, three 533mm (21in) torpedo tubes
Complement:	186
Launched:	1942

Gatling

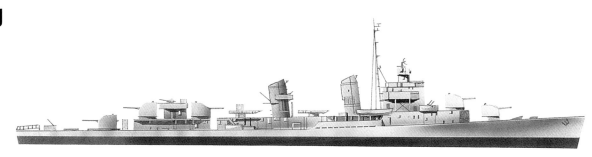

The *Gatling* class were the largest wartime destroyers built for the US Navy, with a displacement about 1000 tonnes greater than most predecessors, and firepower little less than that of some light cruisers. *Gatling* remained in service for over 30 years, being finally stricken from the Navy List in 1974.

SPECIFICATIONS	
Type:	US destroyer
Displacement:	2971 tonnes (2924 tons)
Dimensions:	114.7m x 12m x 4.2m (376ft 5in x 39ft 4in x 13ft 9in)
Machinery:	Twin screws, turbines
Top speed:	35 knots
Main armament:	Five 127mm (5in) guns
Launched:	June 1943

Duncan

One of over 90 *Gearing* class ocean-going destroyers, provided with a powerful armament and sufficient fuel supplies for long-range action, *Duncan* was converted to a radar picket ship in 1945, losing its torpedo tubes. It was stricken from the Navy List in 1973, but several of the class served as late as 1980.

SPECIFICATIONS	
Type:	US destroyer
Displacement:	3606 tonnes (3549 tons)
Dimensions:	120m x 12.4m x 5.8m (390ft 6in x 41ft x 19ft)
Machinery:	Twin screws, geared turbines
Top speed:	33 knots
Main armament:	Six 127mm (5in) guns, ten 533mm (21in) torpedo tubes
Complement:	336
Launched:	October 1944

1943   1944

Fletcher

Fletcher was lead ship of a class of 175, built to plans entirely different to its 1930s predecessors, flush-decked like earlier US destroyers, less top-heavy and so less inclined to roll. *Fletcher* gained 15 batttle stars in World War II and five more in the Korean War. Placed in reserve in 1962, it was stricken in 1967.

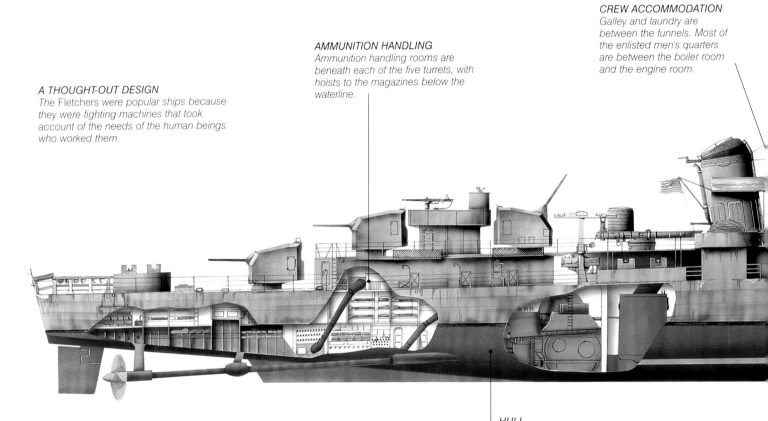

CREW ACCOMMODATION
Galley and laundry are between the funnels. Most of the enlisted men's quarters are between the boiler room and the engine room.

AMMUNITION HANDLING
Ammunition handling rooms are beneath each of the five turrets, with hoists to the magazines below the waterline.

A THOUGHT-OUT DESIGN
The Fletchers were popular ships because they were fighting machines that took account of the needs of the human beings who worked them.

HULL
Displacement was 25 per cent up on the previous Gleaves class, allowing for heavier guns and slightly more crew space.

Fletcher

The *Fletcher* class was the largest class of destroyer ordered, and was also one of the most successful and popular with the destroyer crews themselves. Compared to earlier classes built for the Navy, they carried a significant increase in anti-aircraft (AA) weapons and other weaponry.

SPECIFICATIONS

Type:	US destroyer
Displacement:	2971 tonnes (2924 tons)
Dimensions:	114.7m x 12m x 4.2m (376ft 5in x 39ft 4in x 13ft 9in)
Machinery:	Twin screws, geared turbines
Top speed:	38 knots
Main armament:	Five 127mm (5in) guns, ten 533mm (21in) torpedo tubes
Complement:	273
Launched:	1942

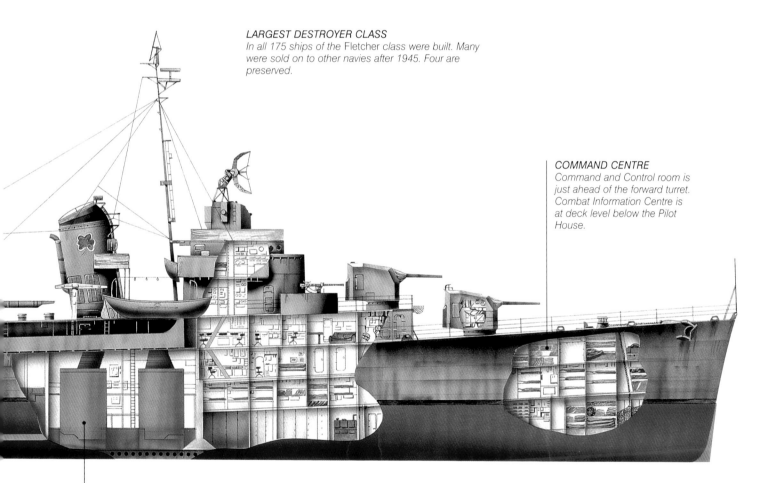

LARGEST DESTROYER CLASS
In all 175 ships of the Fletcher *class were built. Many were sold on to other navies after 1945. Four are preserved.*

COMMAND CENTRE
Command and Control room is just ahead of the forward turret. Combat Information Centre is at deck level below the Pilot House.

MACHINERY
Four Babcock & Wilcox boilers supplying steam to three General Electric geared turbines, with maximum power output of 45MW (60,000 shp).

Destroyers, Part 5: Italian

Five destroyer classes were built by the Italians between 1930 and 1941, none of them numerous, amounting to 36 new ships. Large-sized, they were built very much with an eye on what the French were doing. They were fast, with 38 knots a common speed. The *Soldati* class carried a star-shell gun to help in night actions.

Alberto da Giussano

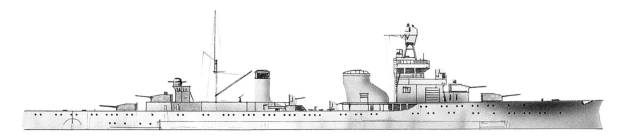

One of four ships built in response to the powerful French *Lion* class destroyers, *Alberto di Giussano* was a large and well-armed vessel, though lightly armoured. It was also very fast – one of the class achieved a speed of 42 knots during trials and maintained a steady 40 knots for eight hours.

SPECIFICATIONS

Type:	Italian fast destroyer
Displacement:	5170 tonnes (5089 tons)
Dimensions:	169.4m x 15.2m x 4.3m (555ft 9in x 49ft 10in x 14ft 1in)
Machinery:	Twin screws, geared turbines
Top speed:	40 knots
Main armament:	Eight 152mm (6in) guns
Launched:	1930

Baleno

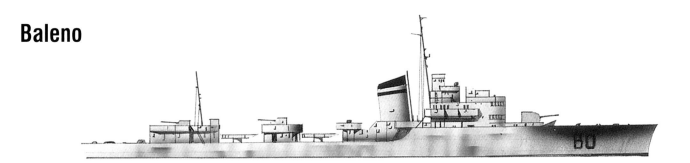

Single-funnelled, with a slimmer hull than its predecessors, *Balena* was sleek and fast. These ships saw hard service in the Mediterranean as convoy escorts, scouts and patrols. On 16 April 1941, it and another destroyer, *Luca Tango*, were in action with four British destroyers. *Baleno* capsized and sank the next day.

SPECIFICATIONS

Type:	Italian destroyer
Displacement:	2123 tonnes (2090 tons)
Dimensions:	94.3m x 9.2m x 3.3m (309ft 6in x 30ft x 10ft 9in)
Machinery:	Twin screws, geared turbines
Top speed:	39 knots
Main armament:	Four 119mm (4.7in) guns, six 533mm (21in) torpedo tubes
Launched:	March 1931

TIMELINE

1930   1931

Fulmine

Destroyer design took a leap forward in 1929–30 with a new Italian class of eight single-funnelled ships with two banks of torpedo tubes placed on the centreline astern of the funnel. *Fulmine* was in the second group. Highly active in World War II, six were lost, including *Fulmine*, sunk in action against British forces.

SPECIFICATIONS	
Type:	Italian destroyer
Displacement:	2124 tonnes (2090 tons)
Dimensions:	94.5m x 9.25m x 3.25m (309ft 6in x 30ft 6in x 11ft)
Machinery:	Twin screws, turbines
Top speed:	38 knots
Main armament:	Four 119mm (4.7in) guns
Launched:	August 1931

Vincenzo Gioberti

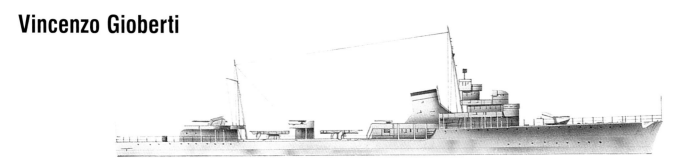

Vincenzo Gioberti was one in a class of four powerful destroyers. The torpedo tubes were mounted on triple carriages placed on the centreline. Later, 20mm (0.8in) anti-aircraft guns were fitted. It was built in 1936–37. On 9 August 1943, it was sunk by a torpedo from the British submarine *Simoom*.

SPECIFICATIONS	
Type:	Italian destroyer
Displacement:	2326 tonnes (2290 tons)
Dimensions:	106.7m x 10m x 3.4m (350ft x 33ft 4in x 11ft 3in)
Machinery:	Twin screws, turbines
Top speed:	39 knots
Main armament:	Four 120mm (4.7in) guns, six 533mm (21in) torpedo tubes
Launched:	September 1936

Artigliere

Artigliere was in the *Soldati* class, the most numerous class of destroyer built for the Italian Navy. All 21 saw extensive war service as escorts able to give and take a deal of punishment. The anti-aircraft defence was inadequate and was soon improved. *Artigliere* was lost in action in October 1940.

SPECIFICATIONS	
Type:	Italian destroyer
Displacement:	2540 tonnes (2500 tons)
Dimensions:	106.7m x 10.2m x 3.5m (350ft x 33ft 4in x 11ft 6in)
Machinery:	Twin screws, geared turbines
Top speed:	38 knots
Main armament:	Four 120mm (4.7in) guns
Launched:	December 1937

1936   1937

Destroyers, Part 6: Italian & Japanese

Japan entered World War II with a total of 110 destroyers, a number equivalent to the American total, deployed in six flotillas. But only 33 more were built. They were efficient craft, highly drilled for night actions when torpedoes were at their most effective, and sharing the Long Lance torpedo with the submarine fleet.

Asashio

The ten *Asashio* class of large destroyers marked Japan's abandonment of treaty limitations on warship construction. Their steam turbines proved unreliable at first, and defects in the steering gear was not corrected until December 1941. All ten ships were lost during World War II, *Asashio* was sunk by US carrier-borne aircraft.

SPECIFICATIONS	
Type:	Japanese destroyer
Displacement:	2367 tonnes (2330 tons)
Dimensions:	118.2m x 10.4m x 3.7m (388ft x 34ft x 12ft)
Machinery:	Twin screws, geared turbines
Top speed:	35 knots
Main armament:	Six 127mm (5in) guns
Complement:	200
Launched:	December 1936

Hamakaze

The first Japanese destroyer with radar, *Hamakaze* and its 17 sisters were armed with six 127mm (5in) guns in twin turrets, but in 1943–44 the upper aft turret was replaced by anti-aircraft guns. The torpedo tubes were positioned amidships in enclosed quadruple mounts. *Hamakaze* was sunk by US aircraft on 7 April 1945.

SPECIFICATIONS	
Type:	Japanese destroyer
Displacement:	2489 tonnes (2450 tons)
Dimensions:	118.5m x 10.8m x 3.7m (388ft 9in x 35ft 5in x 12ft 4in)
Machinery:	Twin screws, turbines
Top speed:	35 knots
Main armament:	Four 127mm (5in) guns, eight 610mm (24in) torpedo tubes
Launched:	November 1940

TIMELINE

1936  1940  1942

Bombardiere

SPECIFICATIONS	
Type:	Italian destroyer
Displacement:	2540 tonnes (2500 tons)
Dimensions:	107m x 10m x 4m (350ft 33ft 7in x 11ft 6in)
Machinery:	Twin screws, turbines
Top speed:	38 knots
Main armament:	Five 120mm (4.7in) guns
Launched:	March 1942

Bombardiere was one of the second group of *Soldati* destroyers, with minor modifications. Some were fitted with depth-charge throwers and all were given additional light AA guns. *Bombardiere* was one of 10 *Soldati* lost in World War II; it was sunk off Marettimo by the British submarine *United* on 17 January 1943.

Ariete

SPECIFICATIONS	
Type:	Italian destroyer
Displacement:	1127.8 tonnes (1110 tons)
Dimensions:	83.5m x 8.6m x 3.15m (274ft x 28ft 3in x 10ft 4in)
Machinery:	Twin screws, geared turbines
Top speed:	31.5 knots
Main armament:	Two 100mm (3.9in) guns, six 450mm (17.7in) torpedo tubes
Complement:	150
Launched:	1943

Built at Trieste, *Ariete* was the lead ship of Italy's final class of World War II destroyers, and the only one of the class to be commissioned before the surrender of Italy in 1943. After World War II, it was transferred to Yugoslavia as part of a war reparations package, and served as *Durmitor* until 1963.

Dragone

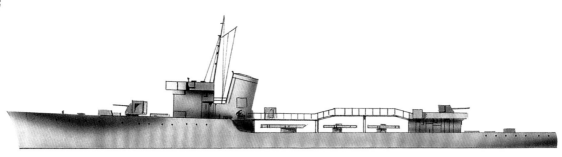

SPECIFICATIONS	
Type:	Italian destroyer/escort
Displacement:	1117 tonnes (1100 tons)
Dimensions:	83.5m x 8.6m x 3m (274ft x 28ft 3in x 10ft 4in)
Machinery:	Twin screws, turbines
Main armament:	Two 102mm (4in) guns, six 450mm (17.7in) torpedo tubes
Launched:	August 1943

Lightweight but slow destroyers, *Dragone* and its sisters were enlargements of the *Spica* class. Built economically, they were vulnerable to faster, more heavily gunned craft. Seized by the Germans after Italy's surrender and numbered TA30, *Dragone* was sunk in June 1944 by torpedoes fired from British MTBs.

1943

Destroyers, Part 7: Japanese & German

Germany's destroyers of the 1930s were advanced ships for their time, but their high-pressure boilers and turbines often failed. Compared to British destroyers they had limited range and magazine capacity. But, numbering only 40, they were a modest part of the *Reich* fleet compared to the hundreds of U-boats.

T1

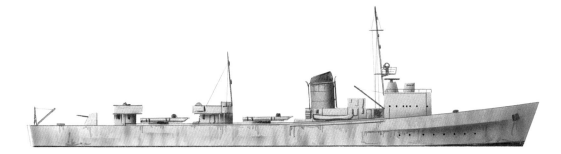

Officially classed as torpedo boats, these were, in fact, small destroyers. There were 36 built for the *Kriegsmarine*, some larger than *T1*, though there was no special function for them. Too small for fleet work, too light for escort duty, their main virtue was speed, making them effective as North Sea raiding vessels.

SPECIFICATIONS

Type:	German light destroyer
Displacement:	1107.4 tonnes (1090 tons)
Dimensions:	84.3m x 8.6m x 2.35m (276ft 7in x 28ft 3in x 7ft 8in)
Machinery:	Twin screws, geared turbines
Top speed:	35 knots
Main armament:	One 105mm (5in) gun, three 533mm (21in) torpedo tubes
Complement:	119
Launched:	1938

Z30

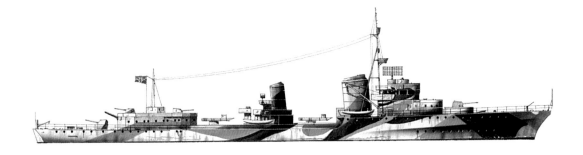

The 'Narvik' class, *Z23–30*, were officially class 36A. Their main guns were the heaviest mounted on destroyers but were not suitable for rapid fire, and their weight reduced the ships' seakeeping qualities. *Z30* was taken by the British at the end of World War II and destroyed in underwater explosive tests.

SPECIFICATIONS

Type:	German destroyer
Displacement:	3750 tonnes (3691 tons) full load
Dimensions:	127m x 12m x 4.6m (416ft 8in x 39ft 4in x 15ft 2in)
Machinery:	Twin screws, geared turbines
Top speed:	38.5 knots
Main Armament:	Four 150mm (5.9in) guns, eight 533mm (21in) torpedo tubes
Complement:	321
Launched:	1940

TIMELINE

1938   1940  1941

Akitsuki

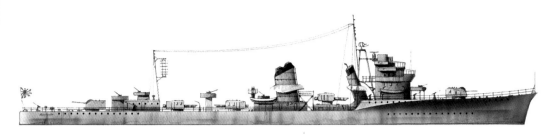

Japan's final class of wartime destroyers carried a newly developed dual-purpose (anti-aircraft, anti-surface craft) gun and long-range torpedoes. *Akitsuki* was sunk by US aircraft at Leyte Gulf in October 1944. Six of its class of 12 survived the war, a high proportion in a fleet almost wholly destroyed.

SPECIFICATIONS	
Type:	Japanese destroyer
Displacement:	3759 tonnes (3700 tons) full load
Dimensions:	134.2m x 11.6m x 4.15m (440ft 3in x 38ft 1in x 13ft)
Machinery:	Twin screws, geared turbines
Top speed:	33 knots
Main armament:	Eight 100mm (3.9in) guns, four 610mm (24in) torpedo tubes
Complement:	300
Launched:	1941

Fuyutsuki

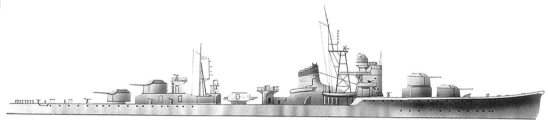

Fuyutsuki was one of a large class of big ocean-going destroyers, for which plans were drawn up in 1939. They were intended as fast anti-aircraft escort ships for the Japanese carrier task forces. By 1943, needs had changed and the design was modified to carry quadruple torpedo tubes for anti-ship attacks.

SPECIFICATIONS	
Type:	Japanese destroyer
Displacement:	3759 tonnes (3700 tons)
Dimensions:	134.2m x 11.6m x 4.2m (440ft 3in x 38ft x 13ft 9in)
Machinery:	Twin screws, turbines
Top speed:	33 knots
Main armament:	Eight 96mm (3.8in) guns, four 607mm (23.9in) torpedo tubes in one quadruple mount
Launched:	January 1944

Fionda

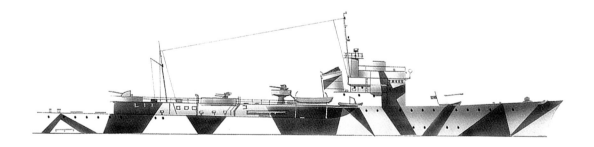

Italian war estimates provided for building 42 boats in *Fionda's* class. Only 16 were laid down, including *Fionda* in 1942. Captured by the Germans on the slips and designated *TA46,* it was damaged in an air raid in 1945. Construction began again, with the new name *Velebit*, but the ship was never completed.

SPECIFICATIONS	
Type:	Italian destroyer
Displacement:	1138 tonnes (1120 tons) approx
Dimensions:	82.2m x 8.6m x 2.8m (270ft x 28ft 3in x 9ft 2in)
Machinery:	Twin screws, turbines
Main armament:	Two 100mm (3.9in) guns
Launched:	Not completed

1944

Karl Galster

Also known as *Z20, Karl Galster* was a considerable improvement on the earlier Type 36, with a new 'clipper' bow shape. Well-armed, with an array of AA guns and carrying 60 mines, it was the only ship of its class to survive World War II. Taken by the Soviets and renamed *Protshnyi*, it was decommissioned around 1960.

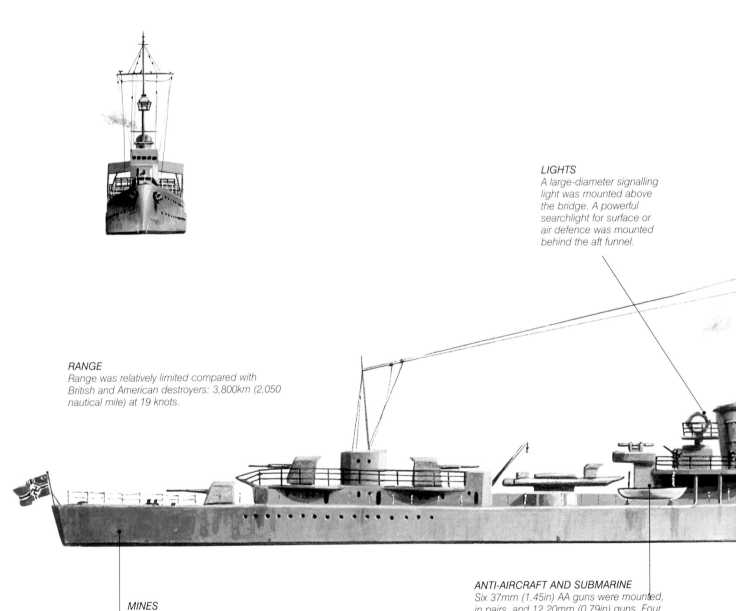

LIGHTS
A large-diameter signalling light was mounted above the bridge. A powerful searchlight for surface or air defence was mounted behind the aft funnel.

RANGE
Range was relatively limited compared with British and American destroyers: 3,800km (2,050 nautical mile) at 19 knots.

MINES
As often with German destroyers, most of the Z20 class were equipped for minelaying, with storage for 60 mines.

ANTI-AIRCRAFT AND SUBMARINE
Six 37mm (1.45in) AA guns were mounted, in pairs, and 12 20mm (0.79in) guns. Four anti-submarine depth-charge launchers were also carried.

Karl Galster

Karl Galster was the only one of its class to survive; the other five were sunk either at Narvik or in Rombaksfjord. In 1946 *Karl Galster* was handed over to the Soviet Navy as part of Germany's war reparations to the Soviet Union. She served in the Baltic fleet as the *Protshnyi* into the 1950s.

SPECIFICATIONS	
Type:	German destroyer
Displacement:	3469.7 tonnes (3415 tons) full load
Dimensions:	125m x 11.8m x 4m (410ft 1in x 38ft 8in x 13ft 1in)
Machinery:	Twin screws, geared turbines
Top speed:	36 knots
Main armament:	Five 127mm (5in) guns, eight 533mm (21in) torpedo tubes
Complement:	330
Launched:	1938

PROFILE
The class is distinguished from its 1934A predecessors by flatter funnel caps, marginally wider beam, and greater length (by 5m, or 15ft).

MACHINERY
The high pressure boilers and turbines, with a substantial power output of 55,554kW (74,500 shp), were more reliable than in previous classes.

BOATS
Four boats were normally carried: captain's gig, cutter, dinghy and Verkehrsboot or general purpose launch, for crew transport and inter-ship trips.

Destroyers, Part 8: German & Other Navies

In the latter years of World War II and into the 1950s, the design and operation of destroyers were influenced by the experience of war. Armament, machinery and arrangement, tested in the ultimate conditions, were relatively unchanged. The main difference with pre-war destroyer types was more effective anti-aircraft defences.

Gravina

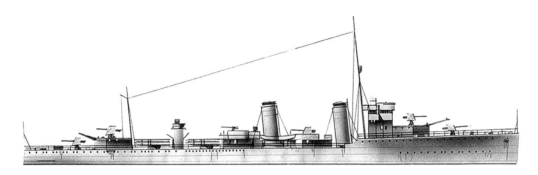

Gravina was one of 16 of the largest destroyers built for the Spanish Navy, virtual copies of the British *Scott* class flotilla leaders. All were launched between 1926 and 1933. The second octet, including *Gravina*, had large gun shields. Range at 14 knots was 8550km (4500 miles). The ship was stricken in the 1960s.

SPECIFICATIONS

Type:	Spanish destroyer
Displacement:	2209 tonnes (2175 tons)
Dimensions:	101.5m x 9.6m x 3.2m (333ft x 31ft 9in x 10ft 6in)
Machinery:	Twin screws, turbines
Top speed:	36 knots
Main armament:	Five 120mm (4.7in) guns, six 533mm (21in) torpedo tubes
Launched:	December 1931

Goteborg

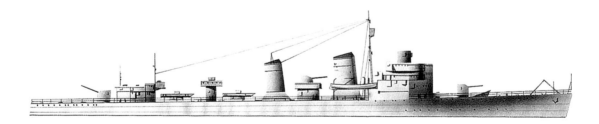

By 1934, Sweden began constructing new destroyers with a class of six to be completed by 1941. The 120mm (4.7in) guns were housed in single mounts, one forward, one aft and one amidships. Sunk by an internal explosion in 1941, then raised, it served until 1958, and was expended as a target in August 1962.

SPECIFICATIONS

Type:	Swedish destroyer
Displacement:	1219 tonnes (1200 tons)
Dimensions:	94.6m x 9m x 3.8m (310ft 4in x 29ft 6in x 12ft 6in)
Machinery:	Twin screws, turbines
Top speed:	42 knots
Main armament:	Three 120mm (4.7in) guns, six 533mm (21in) torpedo tubes
Launched:	October 1935

TIMELINE

1931 　1935 　1938

Vasilefs Georgios

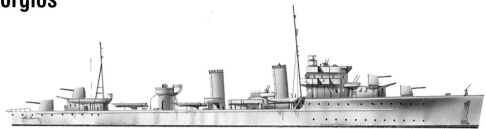

Built for the Greek Navy in Glasgow to the British 'G'-class model, with modified armament, the ship was damaged at Salamis by German bombs. When Greece fell, it was made operational by the Germans as *ZG 3,* later named *Hermes.* Disabled by air attack in April 1943, it was scuttled in the harbour mouth at Tunis.

SPECIFICATIONS	
Type:	Greek destroyer
Displacement:	2123.4 tonnes (2090 tons) full load
Dimensions:	101.2m x 10.4m x 3m (332ft x 34ft 1in x 9ft 10in)
Machinery:	Twin screws, geared turbines
Top speed:	32 knots
Main armament:	Four 127mm (5in) guns, eight 533mm (21in) torpedo tubes
Complement:	215
Launched:	1938

Z51

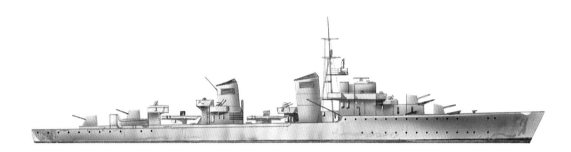

This was intended as a prototype for fleet torpedo boats, more lightly armed than previous German destroyer types but more manoeuvrable, and with a greater array of anti-aircraft weapons. *Z51*, while under construction in Bremen, was sunk by RAF Mosquito bombers in March 1945. No others were built.

SPECIFICATIONS	
Type:	German destroyer
Displacement:	2674.1 tonnes (2632 tons)
Dimensions:	114.3m x 11m x 4m (375ft 8in x 37ft 8in x 13ft 1in)
Machinery:	Triple screws, diesels
Top speed:	36.5 knots
Main armament:	Four 127mm (5in) guns, six 533mm (21in) torpedo tubes
Complement:	235
Launched:	1944

Araguaya

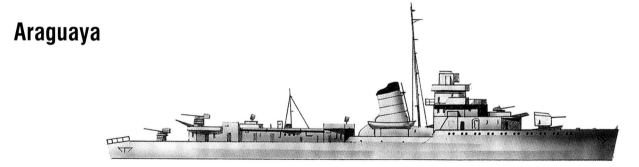

Araguaya and its five sisters were built to replace six Brazilian destroyers taken over by Britain's Royal Navy at the outbreak of World War II. They followed the same original design but used American equipment. All were built between 1943 and 1946 at the Ilha das Cobras Navy Yard. *Araguaya* was discarded in 1974.

SPECIFICATIONS	
Type:	Brazilian destroyer
Displacement:	1829 tonnes (1800 tons)
Dimensions:	98.5m x 10.7m x 2.6m (323ft x 35ft x 8ft 6in)
Machinery:	Twin screws, geared turbines
Top speed:	35.5 knots
Main armament:	Four 127mm (5in) guns, two 40mm (1.57in) guns
Launched:	1946

1944   1946

Gunboats & Motor Boats

Intended for close-quarters and inshore hostilities, the motor gun boat could also be used on short-range coastal raids. The torpedo boat was also motorized, and the Soviets in particular built many in the 1930s. Larger gunboats made longer-range patrols, or were designed for specific locations, including lakes and rivers.

Eritrea

Officially designated as a sloop, *Eritrea* could use its diesel and electric motors independently or together. Maximum range was 9500km (5000 miles) at 15.3 knots. Refitted as a minelayer in 1940–41, it was captured by the British in 1943. Handed to France in 1948, it was renamed *Francis Gamier*, serving until 1966.

SPECIFICATIONS	
Type:	Italian sloop
Displacement:	3117 tonnes (3068 tons)
Dimensions:	96.9m x 13.3m x 4.7m (318ft x 43ft 8in x 15ft 5in)
Machinery:	Twin screws, diesel engines plus electric drive
Top speed:	18 knots
Main armament:	Four 120mm (4.7in) guns
Launched:	September 1936

Dragonfly

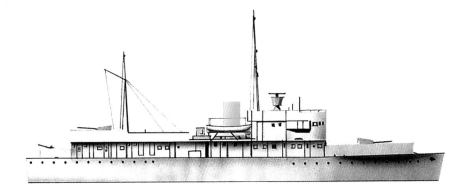

Dragonfly was one of a class of four river gunboats. These were compact with a shallow draught, able to navigate shallow rivers. Engines developed 3800hp and they carried 90 tonnes (90 tons) of fuel oil. *Dragonfly* was sunk by Japanese dive bombers while trying to escape from Singapore on 14 February 1942.

SPECIFICATIONS	
Type:	British gunboat
Displacement:	726 tonnes (715 tons)
Dimensions:	60m x 10m x 1.8m (196ft 6in x 33ft 8in x 6ft 2in)
Machinery:	Twin screws, turbines
Top speed:	17 knots
Main armament:	Two 102mm (4in) guns
Complement:	74
Launched:	1938

TIMELINE   1936   1938

G5

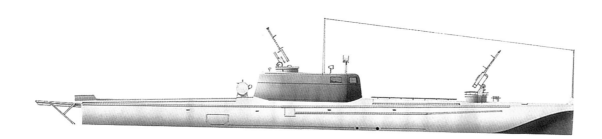

G5 was one in a class of about 295 high-speed vessels, conceived in the early 1930s. Many had Isotta Fraschini engines, which were generally reliable, though Soviet-built versions had a poorer record. Torpedoes were launched from aft, and as they struck the water the vessel had to veer sideways out of their way.

SPECIFICATIONS	
Type:	Soviet torpedo gunboat
Displacement:	14 tonnes (13.7 tons)
Dimensions:	19.1m x 3.4m x 0.75m (62ft 6in x 11ft x 2ft 6in)
Machinery:	Twin screws, petrol engines
Top speed:	45 knots
Main armament:	Two 12.7mm (.5in) guns, two 533mm (21in) torpedo tubes
Launched:	1938

D 3 Type MTB

Around 130 craft of this type were built for the Soviet Navy, mainly for use in the Baltic Sea. The torpedoes were carried in and launched from cradles rather than tubes. Though small compared with their German counterparts, and basic in construction, they were effective boats, able to withstand severe sea conditions.

SPECIFICATIONS	
Type:	Soviet motor torpedo boat
Displacement:	32.5 tonnes (32 tons)
Dimensions:	21.6m x 3.95m x 1.35m (71ft x 13ft x 4ft 6in)
Machinery:	Triple screws, GAM-34FN petrol engines
Top speed:	39 knots
Main armament:	Two 533mm (21in) torpedo cradles
Complement:	9–14
Launched:	1939–45

Fairmile Type C

These wooden-hulled gunboats were enlargements of the RN's Type A motor launch, fitted with Hall-Scott petrol engines and four anti-submarine depth-charges. The type was a failure. The boats had wide turning circles and were exposed while exchanging fire, and noisy motors made clandestine use difficult.

SPECIFICATIONS	
Type:	British motor gunboat
Displacement:	70 tonnes (69 tons)
Dimensions:	33.55m x 6.5m x 1.75m (110ft x 17ft 5in x 5ft 8in)
Machinery:	Triple screws, petrol engines
Top speed:	27 knots
Main armament:	Two 2pdr pom-pom guns, eight machine-guns
Complement:	16
Launched:	Built from 1941

1939   1941

Light Cruisers: Part 1

The London Naval Treaty of 1930 made no distinction between light and heavy cruisers by displacement, but only by gun calibre. A light cruiser was permitted guns of 155mm (6.1in) or less, while a heavy cruiser could be armed with 205mm (8in) guns. The maximum displacement allowed was 10,000 tons (10,106 tonnes).

Arethusa

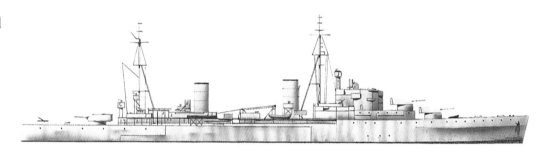

Arethusa and three companion ships were an attempt to create the smallest possible cruiser with reasonable armament and performance. They resembled the slightly longer *Perth* class of 1933. Two of the group were lost in World War II. *Arethusa* was broken up in 1950; one ship, *Aurora*, was sold to China in 1948.

SPECIFICATIONS	
Type:	British cruiser
Displacement:	6822 tonnes (6715 tons)
Dimensions:	154m x 15.5m x 5m (506ft x 51ft x 16ft 6in)
Machinery:	Quadruple screws, geared turbines
Main armament:	Six 152mm (6in) guns, four 102mm (4in) guns
Armour:	51mm (2in) belt
Launched:	1932

Emile Bertin

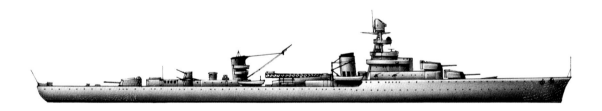

This light cruiser was built to hunt commerce raiders and submarines, with a seaplane. On the fall of France in 1940, it was at Martinique, where it was disarmed. In 1944–45, it was refitted in the United States, with new armament. The aircraft catapult was removed. *Emile Bertin* was scrapped in 1959.

SPECIFICATIONS	
Type:	French cruiser
Displacement:	8615.7 tonnes (8480 tons)
Dimensions:	177m x 16m x 6.6m (580ft 8in x 52ft 6in x 21ft 8in)
Machinery:	Quadruple screws, geared turbines
Top speed:	34 knots
Main armament:	Nine 152mm (6in) guns, six 550mm (21.7in) torpedo tubes
Armour:	25mm (1in) magazines and deck
Complement:	711
Launched:	1933

TIMELINE

1932   1933

Gotland

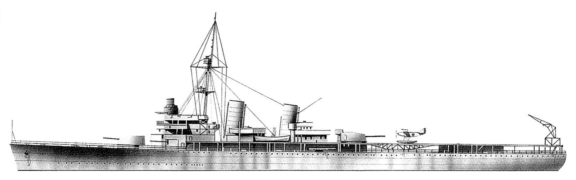

Gotland was planned as a aircraft carrier with 12 float planes, but the design was revised and the designation changed to aircraft cruiser. It usually carried six planes aft, but had room for eight on deck and three below. In 1943–44, it was converted into an anti-aircraft cruiser. It was removed from the list in 1960.

SPECIFICATIONS	
Type:	Swedish cruiser
Displacement:	5638 tonnes (5550 tons)
Dimensions:	134.8m x 15.4m x 5.5m (442ft 3in x 50ft 6in x 18ft)
Machinery:	Twin screws, turbines
Top speed:	28 knots
Main armament:	Six 152mm (6in) guns
Launched:	1933

Tromp

A scout cruiser, *Tromp* escaped to Britain when the Germans occupied Holland and served as an Allied ship throughout World War II, mostly in the Far East. It carried one reconnaissance aircraft. After 1945, it remained in the Dutch Navy until 1958, when it was decommissioned but used as an accommodation ship.

SPECIFICATIONS	
Type:	Dutch cruiser
Displacement:	4337.8 tonnes (4860 tons)
Dimensions:	132m x 12.4m x 4.2m (433ft x 40ft 8in x 13ft 9in)
Machinery:	Twin screws, geared turbines
Top speed:	33.5 knots
Main armament:	Six 150mm (5.9in) guns, six 533mm (21in) torpedo tubes
Armour:	15mm (.7in) belt, 30mm (1.2in) sides, 25mm (1in) deck
Complement:	309
Launched:	1937

Chapayev

Laid down at St Petersburg (then Leningrad), *Chapayev* was not completed until 1950. With a standard displacement of 11,480 tonnes (11,300 tons), it was a large ship to carry only 152mm (6in) guns. The design and armament were virtually obsolete and in 1960 it was disarmed and hulked. It was broken up in 1964.

SPECIFICATIONS	
Type:	Soviet cruiser
Displacement:	15,240 tonnes (15,000 tons) full load
Dimensions:	201m x 19.7m x 6.4m (659ft 5in x 64ft 8in x 21ft)
Machinery:	Twin screws, geared turbines, cruising diesels
Top speed:	34 knots
Main armament:	Twelve 152mm (6in) guns, eight 100mm (3.9in) guns
Complement:	840
Launched:	1940

1937   1940

Light Cruisers: Part 2

Great Britain had by far the largest number of light cruisers, with 81 in commission in 1939, some of them quite elderly. Twenty-three were lost in the course of World War II. The US Navy had 47, Italy had nine, France 12, and the Soviet Union four.

Jacob van Heemskerck

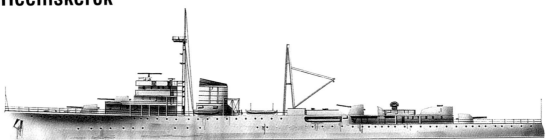

Jacob van Heemskerck was envisaged as a flotilla leader, but the plans were changed in 1938. Then, after the Nazi invasion of Holland in 1939, the ship was taken to Britain for completion and 102mm (4in) guns were mounted, as heavier armament was not available. *Jacob van Heemskerck* was scrapped in 1958.

SPECIFICATIONS	
Type:	Dutch cruiser
Displacement:	4282 tonnes (4215 tons)
Dimensions:	131m x 12m x 4.5m (433ft x 40ft 9in x 15ft)
Machinery:	Twin screws, turbines
Top speed:	34.5 knots
Main armament:	Eight 102mm (4in) guns
Launched:	September 1939

Attilio Regolo

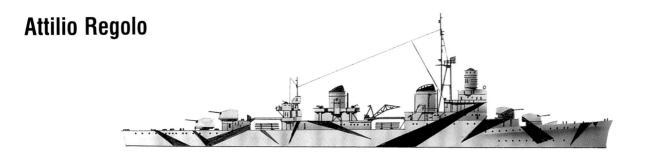

Attilio Regolo was one of 12 fast cruisers of the 'Capitani Romani' class laid down in 1939. Five of the class were not completed, and broken up on the stocks; three were lost in action; and one was scuttled to avoid capture. *Attilio Regolo* was transferred to France in 1948.

SPECIFICATIONS	
Type:	Italian cruiser
Displacement:	5419 tonnes (5334 tons)
Dimensions:	142.9m x 14.4m x 4.9m (469ft x 47ft 3in x 16ft)
Machinery:	Twin screws, turbines
Top speed:	42 knots
Main armament:	Eight 135mm (5.4in) guns
Armour:	20mm (0.8in) over guns
Launched:	August 1940

TIMELINE

1939      1940   1941

Caio Mario

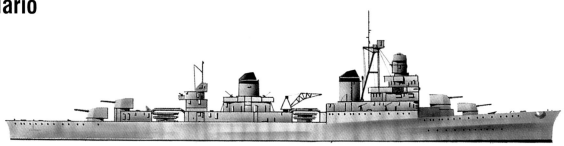

Caio Mario was one of a class of fast cruisers intended as anti-destroyer escorts as well as scouts. High speed was achieved at the expense of protection, and only a splinter-proof deck covered the machinery. It was not yet completed when it was scuttled at La Spezia in 1943 to prevent capture by the Germans.

SPECIFICATIONS	
Type:	Italian cruiser
Displacement:	5419 tonnes (5334 tons)
Dimensions:	143m x 14m x 4.8m (469ft 1in x 46ft x 15ft 7in)
Machinery:	Twin screws, turbines
Top speed:	40 knots
Main armament:	Eight 135mm (5.3in) guns
Launched:	August 1941

Ulpio Traiano

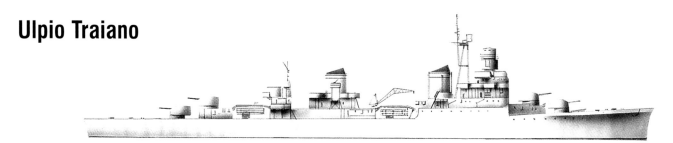

Ulpio Traiano was a sister ship of *Attilio Regolo*. To achieve a high operating speed, protective plating was only 15mm (0.6in) thick on the bridge, and 20mm (0.8in) on the four twin turrets. *Ulpio Traiano* was sunk by British 'human torpedo' commando units while still completing at Palermo harbour in 1943.

SPECIFICATIONS	
Type:	Italian cruiser
Displacement:	5420 tonnes (5334 tons)
Dimensions:	143m x 14.4m x 4.9m (468ft 10in x 47ft 3in x 16ft)
Machinery:	Twin screws, turbines
Top speed:	40 knots
Main armament:	Eight 135mm (5.3in) guns
Launched:	1942

Göta Lejon

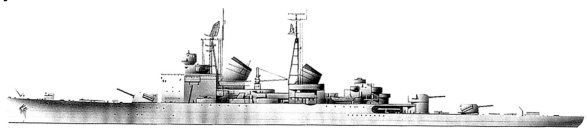

Sweden laid down two well-armed and strongly protected cruisers in 1943. Two raked funnels gave *Göta Lejon* and its sister *Tre Kronor* a distinctive appearance, enhanced on *Göta Lejon* by the tower bridge structure added during rebuilding in 1957–58. *Göta Lejon* was sold to Chile in 1971.

SPECIFICATIONS	
Type:	Swedish cruiser
Displacement:	9347 tonnes (9200 tons)
Dimensions:	182m x 6.7m x 6.5m (597ft x 22ft x 21ft 4in)
Machinery:	Twin screws, turbines
Top speed:	33 knots
Main armament:	Seven 152mm (6in) guns
Launched:	November 1945

1942   1945

Landing Craft & Command Ships

Thousands of landing craft were built between 1942 and 1945 to carry and disembark combat-ready troops, tanks, trucks and support vehicles. Varying in size from 200 to several thousand tonnes, they were a new item in the naval armoury. Supported by escort craft, most also carried light weapons against air attack.

LCI (L)

The Landing Craft Infantry (Large) was designed in the UK as a vessel for raids, carrying up to 188 troops. Over 1000 were built in the United States and Britain. With a range of 15,750km (8,500nm), they could cross the Atlantic. Full load doubled their displacement and gave a stern depth of 1.5m (5ft).

SPECIFICATIONS

Type:	US–British landing craft
Displacement:	197 tonnes (194 tons)
Dimensions:	48.3m x 7.2m x .8m (158ft 6in x 23ft 8in x 2ft 8in)
Machinery:	Twin screws, diesels
Top speed:	15.5 knots
Main armament:	Four 20mm (0.79in) guns
Complement:	24
Launched:	1942–44

Ashland

A self-propelling dry-dock, the LSD pumped its ballast tanks full or dry to lower or raise the stern section for the entrance and exit of landing craft (LCT). Three of these could be fitted, each holding five medium-size tanks. *Ashland* remained on the Navy list until 1970. The concept has been refined in later naval craft.

SPECIFICATIONS

Type:	US landing ship (LSD)
Displacement:	8057 tonnes (7930 tons) full load
Dimensions:	139.5m x 22m x 4.8m (457ft 9in x 72ft 2in x 15ft)
Machinery:	Twin screws, triple expansion reciprocating engines
Top speed:	15.5 knots
Main armament:	One 127mm (5in) gun, twelve 40mm (1.57in) and sixteen 20mm (0.79in) guns
Complement:	254
Launched:	1942

TIMELINE

1942   1943

Appalachian

Appalachian acted as a headquarters vessel ensuring communication and coordination in the large-scale amphibious assaults on Japanese-held islands during World War II. It served briefly as Pacific Fleet flagship in 1947, before being removed from the active list that year. It was broken up in 1960.

SPECIFICATIONS	
Type:	US command ship
Displacement:	14,133 tonnes (13,910 tons)
Dimensions:	132.6m x 19.2m x 7.3m (435ft x 63ft x 24ft)
Machinery:	Single-screw, turbine
Speed:	17 knots
Main armament:	Two 127mm (5in) guns, eight 40mm (1.57in) guns
Launched:	1943

LSM (R)

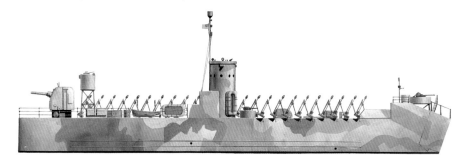

The R variant of the Landing Ship Medium was a fire-support ship, bristling with guns, mortars and rocket launchers to give fire support during troop landings. Dimensions and specifications varied during construction. The earlier ships had a greater operating range, though all could make ocean passages.

SPECIFICATIONS	
Type:	US landing craft/fire support ship
Displacement:	795.5 tonnes (783 tons)
Dimensions:	62m x 10.5m x 1.68m (203ft 6in x 34ft 6in x 5ft 6in)
Machinery:	Twin screws, diesels
Top speed:	13 knots
Main armament:	One 127mm (5in) gun, four 108mm (4.2in) mortars, rockets, light AA guns
Complement:	143
Launched:	Built 1944–45

T1

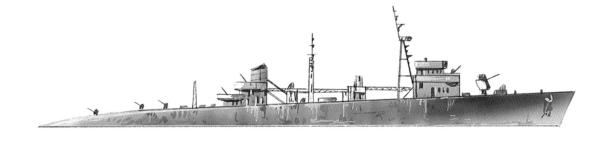

Instead of the dock-type or crane-fitted landing ship, the Japanese produced this type, able to launch five fully-loaded landing craft from rails running off the specially shaped stern. Some of the 22 built were modified in 1945 to launch cradle-mounted midget submarines. Only a few *T1*s survived the war.

SPECIFICATIONS	
Type:	Japanese landing-ship
Displacement:	2235.2 tonnes (2200 tons) full load
Dimensions:	96m x 10.2m x 3.6m (315ft x 33ft 5in x 11ft 10in)
Machinery:	Single screw, geared turbines
Top speed:	22 knots
Main armament:	Two 127mm (5in) guns
Launched:	1944

1944

Minelayers

Estimates of the number of mines laid between 1939 and 1945 range from 600,000 to over a million. Aircraft dropped mines but the majority were laid from ships. The classic minelayer was a fast ship, purpose-built, but many ships were adapted for the purpose, and destroyers often carried a stock of mines.

Gouden Leeuw

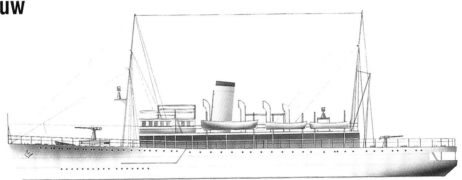

SPECIFICATIONS

Type:	Dutch minelayer
Displacement:	1311 tonnes (1291 tons)
Dimensions:	65.8m x 11m x 3.3m (216ft x 36ft x 11ft)
Machinery:	Twin screws, triple expansion engines
Top speed:	15 knots
Main armament:	Two 76mm (3in) anti-aircraft guns
Launched:	1931

Built for service in the Far East, *Gouden Leeuw* could carry up to 250 mines, depending on size and type, held in a magazine aft, and laid from rails that led to the stern. It was caught and sunk by Japanese warships off Tarakan on 12 January 1942, less than a month after the start of the war in the Pacific.

Gryf

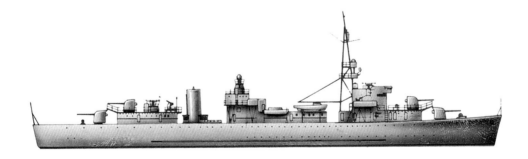

SPECIFICATIONS

Type:	Polish minelayer
Displacement:	2286 tonnes (2250 tons)
Dimensions:	103.2m x 13.1m x 3.6m (338ft 7in x 43ft x 11ft 10in)
Machinery:	Twin screws, diesels
Top speed:	20 knots
Main armament:	Six 120mm (4.7in) guns
Complement:	205
Launched:	1936

Gryf was primarily a minelayer carrying 600 mines, though it was also intended to function as a training vessel and as a state yacht. Built in France, delivered in 1938, it was deliberately sunk in a floating dry-dock at Hela, as a defensive battery, on 1 September 1939. On the 3rd it was destroyed by German aircraft.

TIMELINE

1931 1936 1940

Tsugaru

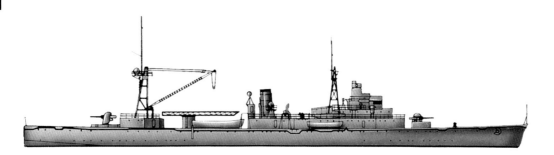

Comparable in size to the British *Abdiel* class (see *Ariadne*), this was one of only two larger Japanese minelayers. As with the British ships, the mines were held on a deck running almost full length, and dropped from the stern. *Tsugaru* also carried an aircraft. It was torpedoed by a US submarine in June 1944.

SPECIFICATIONS	
Type:	Japanese minelayer
Displacement:	6705.6 tonnes (6600 tons) full load
Dimensions:	124.5m x 15.6m x 4.9m (408ft 6in x 51ft 3in x 16ft 2in)
Machinery:	Twin screws, geared turbines
Top speed:	20 knots
Main armament:	Four 127mm (5in) guns
Launched:	1940

Artevelde

This multi-purpose vessel was also intended to operate as a fishery patrol ship. Captured while still building in May 1940, *Artevelde* was completed by the Germans and renamed *Lorelei*. At the end of World War II, it was returned to Belgium, where it served until the early 1950s. It was broken up in 1954–55.

SPECIFICATIONS	
Type:	Belgian minelayer and royal yacht
Displacement:	2306 tonnes (2270 tons)
Dimensions:	98.5m x 10.5m x 3.3m (323ft 2in x 34ft 5in x 10ft 10in)
Machinery:	Twin screws, geared turbines
Top speed:	28.5 knots
Main armament:	Four 104mm (4.1in) guns, 120 mines
Launched:	1940

Ariadne

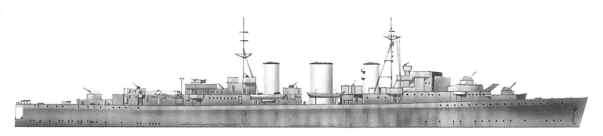

Ariadne was one of six fast minelaying cruisers. They had a small silhouette, and good armament against both air and surface attack. All saw extensive war service, including hazardous ammunition runs to Malta, using their high speed. Three were sunk through enemy action. *Ariadne* was broken up in 1965.

SPECIFICATIONS	
Type:	British cruiser/minelayer
Displacement:	4064 tonnes (4000 tons)
Dimensions:	127.4m x 12.2m x 4.5m (418ft x 40ft x 14ft 9in)
Machinery:	Twin screws, turbines
Top speed:	40 knots
Main armament:	Six 102mm (4in) guns, 100–156 mines
Launched:	1943

1943

Minesweepers

Between 600,000 and a million mines were laid in World War II. Magnetic mines, invented in Germany, replaced contact mines that exploded on touch. The Allies had to find new ways to deal with them. Instead of cutting wires, minesweepers towed electrically-charged cables between two ships. Their task remained hazardous.

Espiègle

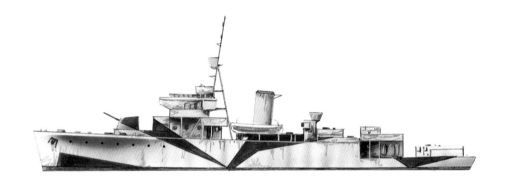

One of the *Algerine* class of minesweepers, *Espiègle* was among relatively few to have turbine engines. Early minesweepers had often been converted trawlers, but this was a more substantial and capable ship. With a range of 22,000km (12,000 miles) at 12 knots, it could go further and stay at sea much longer.

SPECIFICATIONS	
Type:	British minesweeper
Displacement:	995.7 tonnes (980 tons)
Dimensions:	65.6m x 10.8m x 3.2m (225ft x 35ft 6in x 10ft 6in)
Machinery:	Twin screws, turbines
Top speed:	16.5 knots
Main Armament:	One 102mm (4in) gun
Complement:	85
Launched:	1942

YMS 100

The Yard Mine Sweepers were wooden-hulled small warships intended for coastal service. Even so, many made trans-ocean voyages to act as advance way-clearers for invasion forces far from the United States. A submarine-chaser version was tried, but the diesel motors could not produce the necessary speed.

SPECIFICATIONS	
Type:	US minesweeper
Displacement:	365.7 tonnes (360 tons)
Dimensions:	41.45m x 7.45m x 2.35m (136ft x 24ft 6in x 7ft 9in)
Machinery:	Twin screws, diesels
Top speed:	15 knots
Main armament:	One 76mm (3in) gun
Complement:	60
Launched:	Built 1942–44

TIMELINE

1942   1943

T 371

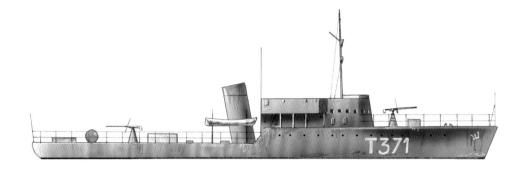

The 'T' class minesweepers, intended for inshore work, were war-production vessels, built quickly and economically from welded flat steel sections and driven by modified tank engines. Protection was minimal. About 145 had been launched by the end of World War II, and another 100 or so were built later.

SPECIFICATIONS

Type:	Soviet minesweeper
Displacement:	152.4 tonnes (150 tons)
Dimensions:	39m x 5m x 1.5m (127ft 11in x 18ft x 4ft 11in)
Machinery:	Twin screws, diesels
Top speed:	14 knots
Main Armament:	Two 45mm (1.8in) guns
Complement:	32
Launched:	1943

Daino

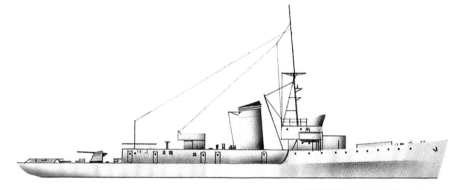

Built of prefabricated sections as the German minesweeper *B2* (later *M802*), this served with the German Minesweeping Administration in the North and Baltic seas at the end of World War II. Transferred to Italy in 1949 as *Daino*, it was a minesweeper, then an escort ship. In 1960 it became an unarmed survey vessel.

SPECIFICATIONS

Type:	Italian minesweeper
Displacement:	850 tonnes (838 tons)
Dimensions:	68m x 9m x 2m (224ft x 29ft 6in x 7ft 3in)
Machinery:	Twin screws, triple expansion engines
Top speed:	14 knots
Launched:	1945

Guadiaro

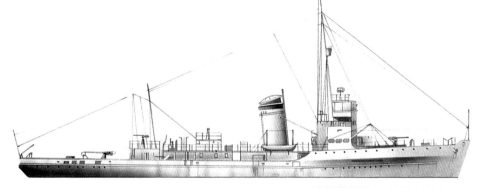

Guadiaro was in the first group of minesweepers built for the Spanish Navy after World War II. The design was modelled on the successful *Bidasoa*, launched in 1943 and built to the plan of the coal-fired German M1940 sweepers. *Guadiaro* was modernized between 1959 and 1960, and withdrawn from service by 1980.

SPECIFICATIONS

Type:	Spanish minesweeper
Displacement:	782 tonnes (770 tons)
Dimensions:	74.3m x 10.2m x 3.7m (243ft 9in x 33ft 6in x 12ft)
Machinery:	Twin screws, triple expansion engines plus exhaust turbines
Top speed:	16 knots
Main armament:	Two 20mm (0.79in) anti-aircraft guns
Launched:	June 1950

1945    1950

Repair, Supply & Support Ships

The increasing scale of fleets, especially in submarines and destroyers, required 'mother ships' to supply food, stores and fuel. The complexity of air-sea combined operations, and the advance of radar and other wireless technology, also brought about the command ship, a centre for co-ordination and communication.

Gustave Zédé

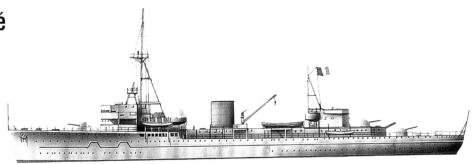

The German submarine depot ship *Saar* was acquired by France in 1947, and commissioned into the French Navy in 1949 as *Gustave Zédé*. During the 1960s, it was France's only ship to carry a fully comprehensive command system. By 1967, the vessel had become the flagship of the Fleet Training Centre.

SPECIFICATIONS	
Type:	French command ship
Displacement:	3282 tonnes (3230 tons)
Dimensions:	93.8m x 13.5m x 4.2m (308ft x 44ft 3in x 14ft)
Machinery:	Twin screws, diesel engines
Top speed:	16 knots
Main armament:	Three 104mm (4.1in) guns
Launched:	April 1934

Togo

Togo was a merchant vessel, equipped as an armed raider, *Coronel*, but it failed to break out through the English Channel. With its original name, it was refitted to operate Freya and Würzburg radar, and radio communications systems. Stationed in the Baltic Sea, it was later used to carry troops, and then refugees.

SPECIFICATIONS	
Type:	German fighter direction ship
Displacement:	12,903 tonnes (12,700 tons)
Dimensions:	134m x 17.9m x 7.9m (439ft 7in x 58ft 9in x 25ft)
Machinery:	Single screw, two-stroke double-acting diesel engine
Top speed:	16 knots
Main armament:	Three 105mm (4.1in) guns, two 40mm (1.57in) guns
Launched:	1940

TIMELINE

1934  1940

Fulton

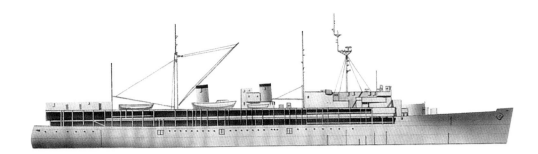

SPECIFICATIONS	
Type:	American submarine tender
Displacement:	18,288 tonnes (18,000 tons)
Dimensions:	161.5m x 22.4m x 7.8m (529ft 10in x 73ft 6in x 25ft 7in)
Machinery:	Twin screws, diesel electric engines
Top speed:	15 knots
Launched:	December 1940

One of a class of seven ships, *Fulton* established seaplane bases in the Panama Canal in December 1941, a prelude to the setting-up of a defence zone around the canal. It served as a submarine tender at Midway during 1942. In the 1950s, its facilities were updated to enable it to support nuclear attack submarines.

Jason

SPECIFICATIONS	
Type:	US repair ship
Displacement:	16,418 tonnes (16,160 tons)
Dimensions:	161.3m x 22.3m x 7m (529ft 2in x 73ft 2in x 23ft 4in)
Machinery:	Twin screws, turbines
Top speed:	19.2 knots
Main armament:	Four 127mm (5in) guns
Complement:	1336
Launched:	December 1940

In 1938, the US Navy authorized *Jason* as a purpose-built repair vessel, with three similar ships to follow. Capable of a multitude of repair and maintenance tasks, it could serve several major surface vessels at once. By the 1980s, the 127mm (5in) guns had been replaced with four 20mm (0.79in) guns.

Norton Sound

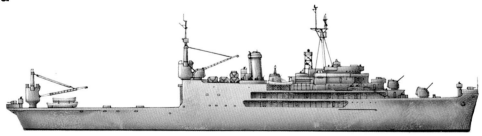

SPECIFICATIONS	
Type:	US seaplane tender
Displacement:	15,341.6 tonnes (15,100 tons)
Dimensions:	164.7m x 21.1m x 6.8m (540ft 5in x 69ft 3in x 22ft 3in)
Machinery:	Twin screws, geared turbines
Top speed:	19 knots
Main armament:	Four 127mm (5in) guns
Complement:	1247
Launched:	1943

One of four *Currituck*-class tenders, commissioned in 1944, *Norton Sound* was built to warship standards. Its after-deck mounted an H-5 hydraulic catapult. Most seaplane tenders, built on merchant ship hulls, were discarded after 1945, but this ship served into the 1980s, latterly as a guided-missile trials ship.

1943

Submarines of the 1930s: Part 1

Big-gun submarines were banned by treaty in 1930, so designers focused their attention on the submarine as a torpedo vehicle. Engine improvements were also pursued, and in the United States lightweight, high-power diesel engines were developed, affording more space for fuel, spare torpedoes and the crew.

Nautilus

Nautilus was one of three V-class submarines. These were slow to dive due to their size and shape – a raised centre deck to improve the guns' field of fire. Re-numbered *SS16* in 1931, *Nautilus* was refitted in 1940 to carry 5104 litres (19,320 gallons) of aviation fuel for seaplanes. It was scrapped in 1945.

SPECIFICATIONS

Type:	US submarine
Displacement:	2773 tonnes (2730 tons) [surface], 3962 tonnes (3900 tons) [submerged]
Dimensions:	113m x 10m (370ft x 33ft 3in)
Machinery:	Twin screws, diesel engines [surface], electric motors [submerged]
Top speed:	17 knots [surface], 8 knots [submerged]
Main armament:	Two 152mm (6in) guns, six 533mm (21in) torpedo tubes
Launched:	March 1930

Delfino

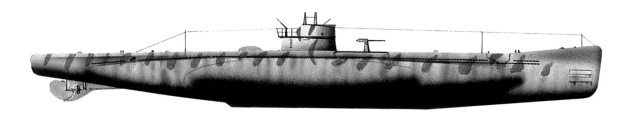

Delfino was completed in 1931. A long-range boat, it was equipped with a gun of sufficient calibre to sink merchant ships when surfaced. A light anti-aircraft gun was mounted on the rear platform of the conning tower. From 1942, *Delfino* was used for training and transport duties. It was sunk in 1943.

SPECIFICATIONS

Type:	Italian submarine
Displacement:	948 tonnes (933 tons) [surface], 1160 tonnes (1142 tons) [submerged]
Dimensions:	70m x 7m x 7m (229ft x 23ft 7in x 23ft 7in)
Machinery:	Twin screws, diesel engines [surface], electric motors [submerged]
Top speed:	15 knots [surface], 8 knots [submerged]
Main armament:	Eight 533mm (21in) torpedo tubes, one 102mm (4in) gun
Launched:	April 1930

TIMELINE

1930   1932

Dolphin

Dolphin was an experimental boat, but the attempt to pack too many features from bigger craft into a relatively small hull was not a success, and the US Navy did not develop a class from it. During World War II, it was used for crew training duties. It was broken up in 1946.

SPECIFICATIONS	
Type:	US submarine
Displacement:	1585 tonnes (1560 tons) [surface], 2275 tonnes (2240 tons) [submerged]
Dimensions:	97m x 8.5m x 4m (319ft 3in x 27ft 9in x 13ft 3in)
Machinery:	Twin screws, diesel engines [surface], electric motors [submerged]
Top speed:	17 knots [surface], 18 knots [submerged]
Main armament:	Six 533mm (21in) torpedo tubes, one 102mm (4in) gun
Launched:	March 1932

U-2

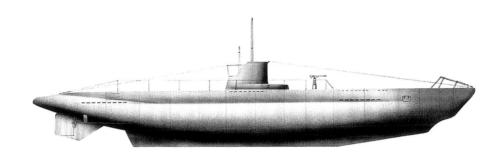

Forbidden to build or possess submarines after World War I, Germany set up clandestine design teams in Spain, Holland and the Soviet Union in the 1920s. A first boat built for Finland in 1927 was the basis for *U-2*, intended for coastal service. All early Type II boats were used for training. *U-2* was sunk in April 1944.

SPECIFICATIONS	
Type:	German submarine
Displacement:	254 tonnes (250 tons) [surface], 302 tonnes (298 tons) [submerged]
Dimensions:	40.9m x 4.1m x 3.8m (133ft 2in x 13ft 5in x 12ft 6in)
Machinery:	Twin screws, diesel engines [surface], electric motors [submerged]
Top speed:	13 knots [surface], 7 knots [submerged]
Main armament:	Three 533mm (21in) torpedo tubes, one 20mm (0.79in) gun
Launched:	July 1935

Enrico Tazzoli

Enrico Tazzoli took part in the Spanish Civil War, and served in the Mediterranean for the early part of World War II. In 1940, it transferred to the Atlantic. In 1942, it was refitted to transport supplies to Japan, but on its first voyage in 1943 it disappeared without trace in the Bay of Biscay.

SPECIFICATIONS	
Type:	Italian submarine
Displacement:	1574 tonnes (1550 tons) [surface], 2092 tonnes (2060 tons) [submerged]
Dimensions:	84.3m x 7.7m x 5.2m (276ft 6in x 25ft 3in x 17ft)
Machinery:	Twin screws, diesel engines [surface], electric motors [submerged]
Top speed:	17 knots [surface], 8 knots [submerged]
Main armament:	Eight 533mm (21mm) torpedo tubes, two 120mm (4.7in) guns
Launched:	October 1935

1935

Submarines of the 1930s: Part 2

Italy built over 100 submarines in the 1930s and many served, sometimes covertly, on behalf of Franco's Nationalist forces in the Spanish Civil War (1936–39). These were mainly short-range boats, as were most British submarines of the era. The Americans were more concerned with submarines capable of long-range cruising.

Aradam

Aradam was one of a class of 17 sturdy short-range vessels. All operated in the Mediterranean Sea during World War II, except *Macalle*, which served in the Red Sea. *Aradam* was scuttled in September 1943 in Genoa harbour to avoid capture. Raised by the Germans, it was sunk by bombing the following year.

SPECIFICATIONS

Type:	Italian submarine
Displacement:	691 tonnes (680 tons) [surface], 880 tonnes (866 tons) [submerged]
Dimensions:	60.2m x 6.5m x 4.6m (197ft 6in x 21ft 4in x 15ft)
Machinery:	Twin screws, diesel engines [surface], electric [submerged]
Top speed:	14 knots [surface], 7 knots [submerged]
Main armament:	Six 530mm (21in) torpedo tubes, one 100mm (4in) gun
Launched:	1936

Diaspro

Diaspro was one of 10 short-range boats of the *Perla* class, designed for use in Mediterranean waters, with a maximum operational depth of 70–80m (230–262ft). All were engaged in the Spanish Civil War, on Spain's east coast, and served in World War II. *Diaspro* was removed from the Navy List in 1948.

SPECIFICATIONS

Type:	Italian submarine
Displacement:	711 tonnes (700 tons) [surface], 873 tonnes (860 tons) [submerged]
Dimensions:	60m x 6.4m x 4.6m (197ft 5in x 21ft 2in x 15ft)
Machinery:	Twin screws, diesel engines [surface], electric motors [submerged]
Top speed:	14 knots [surface], 8 knots [submerged]
Main armament:	Six 533mm (21in) torpedo tubes, one 100mm (3.9in) gun
Launched:	July 1936

TIMELINE

1936

Corallo

Corallo was a short-range submarine of the *Perla* class, all of which served in the Spanish Civil War, where two were ceded to the Nationalist forces. They also served in World War II, and one sank the British cruiser *Bonaventure*. Five were lost, including *Corallo*, sunk in December 1942 by the British sloop *Enchantress*.

SPECIFICATIONS	
Type:	Italian submarine
Displacement:	707 tonnes (696 tons) [surface], 865 tonnes (852 tons) [submerged]
Dimensions:	60m x 6.5m x 5m (197ft 5in x 21ft 2in x 15ft 5in)
Machinery:	Twin screws, diesel engines [surface], electric motors [submerged]
Top speed:	14 knots [surface], 8 knots [submerged]
Main armament:	Six 533mm (21in) torpedo tubes, one 100mm (3.9in) gun
Launched:	August 1936

Dagabur

Dagabur was one of a class of 17. While serving the Nationalists in the Spanish Civil War, two boats were modified to carry small assault craft. These were later used to inflict serious damage on the British vessels *Valiant* and *Queen Elizabeth* in 1941. *Dagabur* was sunk by the British destroyer *Wolverine* in 1942.

SPECIFICATIONS	
Type:	Italian submarine
Displacement:	690 tonnes (680 tons) [surface], 861 tonnes (848 tons) [submerged]
Dimensions:	60m x 6.5m x 4m (197ft 6in x 21ft x 13ft)
Machinery:	Twin screws, diesel engines [surface], electric motors [submerged]
Top speed:	14 knots [surface], 8 knots [submerged]
Main armament:	Six 533mm (21in) torpedo tubes, one 100mm (3.9in) gun
Launched:	November 1936

U-32

U-32 was one of the early Type VII ocean-going submarines, and the basis for all later construction. They were compact, relatively cheap and simple to build, easy to operate and reliable. Between 1941 and 1943, their success in sinking merchant shipping almost defeated Britain. *U-32* was sunk in October 1940.

SPECIFICATIONS	
Type:	German submarine
Displacement:	626 tonnes (616 tons) [surface], 745 tonnes (733 tons) [submerged]
Dimensions:	64.5m x 5.8m x 4.4m (211ft 8in x 19ft x 14ft 5in)
Machinery:	Twin screws, diesel engines [surface], electric motors [submerged]
Top speed:	16 knots [surface], 8 knots [submerged]
Main armament:	Five 533mm (21in) torpedo tubes, one 88mm (3.5in) gun
Launched:	April 1937

1937

Submarines of the 1930s: Part 3

In 1939, the US submarine *Squalus* and the British *Thetis* both accidentally flooded within days of each other, and many lives were lost. The two disasters led to greater attention to safety on board, with interlocks to prevent doors being inadvertently opened, and escape chambers to allow crew members to get out in an emergency.

Dandolo

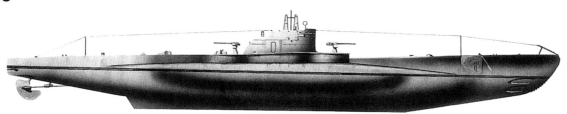

The *Dandolo*s, single-hull boats with internal ballast tanks, were among Italy's best submarines of World War II. Surface range at 17 knots was 4750km (2500 miles); submerged at 9 knots it was 14,250 kilometres (7500 miles). Maximum operational depth was 100m (328ft). All were lost in action except *Dandolo*.

SPECIFICATIONS	
Type:	Italian submarine
Displacement:	1080 tonnes (1063 tons) [surface], 1338 tonnes (1317 tons) [submerged]
Dimensions:	73m x 7.2m x 5m (239ft 6in x 23ft 8in x 16ft 5in)
Machinery:	Twin screws, diesel engines [surface], electric motors [submerged]
Top speed:	17.4 knots [surface], 8 knots [submerged]
Main armament:	Eight 533mm (21in) torpedo tubes, two 100mm (3.9in) guns
Launched:	November 1937

Brin

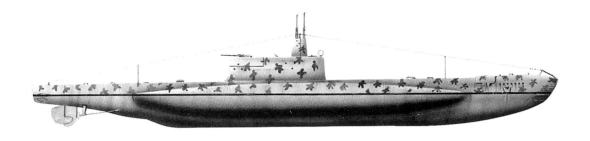

A long-range boat, *Brin* had a partial double hull developed from the *Archimede* class. Early in World War II, it served in the Mediterranean and Atlantic. In 1943 it was stationed in Ceylon, and after the Italian surrender, helped to train British anti-submarine units in the Indian Ocean.

SPECIFICATIONS	
Type:	Italian submarine
Displacement:	1032 tonnes (1016 tons) [surface], 1286 tonnes (1266 tons) [submerged]
Dimensions:	70m x 7m x 4.2m (231ft 4in x 22ft 6in x 13ft 6in)
Machinery:	Twin screws, diesel engines [surface], two electric motors [submerged]
Top speed:	17 knots [surface], 8 knots [submerged]
Main armament:	Eight 533mm (21in) torpedo tubes
Launched:	1938

TIMELINE

1937   1938

Durbo

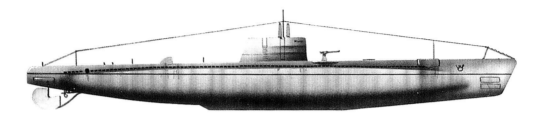

All 17 vessels in the 600 Series *Adua* class, including *Durbo*, gave good service in World War II, but only one survived it. In 1940, *Durbo* was caught on a patrol close to Gibraltar by the British destroyers *Firedrake* and *Wrestler*. Forced to the surface by depth charges, it was scuttled by its crew.

SPECIFICATIONS	
Type:	Italian submarine
Displacement:	710 tonnes (698 tons) [surface], 880 tonnes (866 tons) [submerged]
Dimensions:	60m x 6.4m x 4m (197ft 6in x 21ft x 3ft)
Machinery:	Twin screws, diesel engines [surface], electric motors [submerged]
Top speed:	14 knots [surface], 7.5 knots [submerged]
Main armament:	Six 533mm (21in) torpedo tubes, one 100mm (3.9in) gun
Launched:	March 1938

Galvani

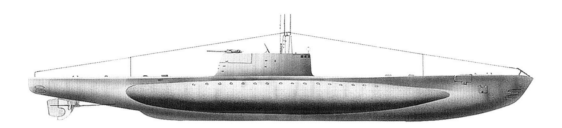

Galvani was one of a group of four long-range submarines, developed from the *Archimede* class. Attached to the Red Sea Flotilla, based at Massawa and other ports on the Ethiopian and Somali coast, it sank the British sloop *Pathan* on 23 June 1940, but was itself was sunk on the following day by HMS *Falmouth*.

SPECIFICATIONS	
Type:	Italian submarine
Displacement:	1032 tonnes (1016 tons) [surface], 1286 tonnes (1266 tons) [submerged]
Dimensions:	72.4m x 6.9m x 4.5m (237ft 6in x 22ft x 14ft 11in)
Machinery:	Twin screws, diesel engines [surface], electric motors [submerged]
Top speed:	17.3 knots [surface], 8 knots [submerged]
Main armament:	Eight 533mm (21in) torpedo tubes, one 99mm (3.9in) gun
Launched:	May 1938

Thistle

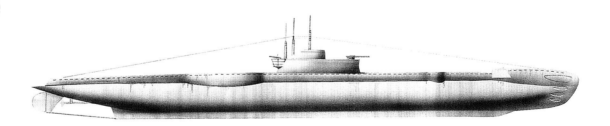

Thistle was one of the first of 21 British 'T'-class submarines, built to replace the previous 'O', 'P' and 'R' classes of ocean-going submarines, and designed to remain at sea for patrols of up to 42 days. *Thistle* was lost in the North Sea on 14 April 1940. Some of the 'T' class continued to serve until 1963.

SPECIFICATIONS	
Type:	British submarine
Displacement:	1347 tonnes (1326 tons) [surface], 1547 tonnes (1523 tons) [submerged]
Dimensions:	83.6m x 8m x 3.6m (274ft 3in x 26ft 6in x 12ft)
Machinery:	Twin screws, diesel engines [surface], electric motors [submerged]
Top speed:	15.25 knots [surface], 9 knots [submerged]
Main armament:	Ten 533mm (21in) torpedo tubes, one 102mm (4in) gun
Launched:	1939

1939

World War II Submarines: Part 1

In 1939, Germany had 57 submarines, of which 38 were considered serviceable. Britain had 58, Italy had 105, and the Soviet Union had 218. In the course of the war, the British total would reach 270, with 73 lost in action; the Soviets built another 54 and lost 109; Germany built over 800 and lost 812.

Barbarigo

Barbarigo's short range, 1425km (750 miles) on the surface, or 228km (120 miles) submerged at three knots, was adequate for Mediterranean operations. In 1943 it was converted into a submersible transport but was spotted on the surface in the Bay of Biscay by Allied aircraft, which sank her in June 1943.

SPECIFICATIONS

Type:	Italian submarine
Displacement:	1059 tonnes (1043 tons) [surface], 1310 tonnes (1290 tons) [submerged]
Dimensions:	73m x 7m x 5m (239ft 6in x 23ft x 16ft 6in)
Machinery:	Twin screws, diesel engines [surface], electric motors [submerged]
Top speed:	17.4 knots [surface], 8 knots [submerged]
Main armament:	Eight 533mm (21in) torpedo tubes; two 100 mm (3.9 in)/47 guns four 13.2 mm (0.52 in) machine guns
Launched:	1938

Archimede

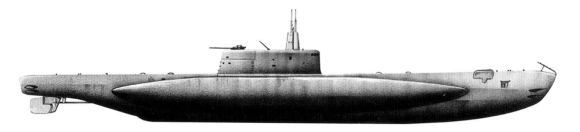

The 100mm (3.9in) conning tower gun was replaced by a 120mm (4.7in) weapon on the foredeck. At the outbreak of World War II, *Archimede* was operating in the Red Sea and the Indian Ocean, then in May 1941 transferred to the Atlantic. It was sunk by Allied aircraft off the Brazilian coast on 14 April 1943.

SPECIFICATIONS

Type:	Italian submarine
Displacement:	1032 tonnes (1016 tons) [surface], 1286 tonnes (1266 tons) [submerged]
Dimensions:	72.4m x 6.7m x 4.5m (237ft 6in x 22ft x 5ft)
Machinery:	Twin screws, diesel engines [surface], electric motor [submerged]
Top speed:	17 knots [surface], 8 knots [submerged]
Main armament:	Eight 533mm (21in) torpedo tubes, one 100mm (3.9in) gun
Launched:	March 1939

TIMELINE

1938   1939   1941

Bronzo

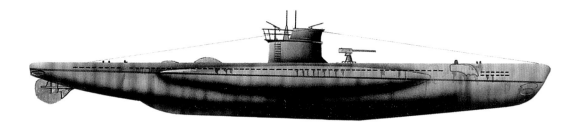

SPECIFICATIONS	
Type:	Italian submarine
Displacement:	726 tonnes (715 tons) [surface], 884 tonnes (870 tons) [submerged]
Dimensions:	60m x 6.5m x 4.5m (197ft 4in x 21ft 4in x 14ft 9in)
Machinery:	Twin screws, diesels on surface, electric motors submerged
Top speed:	15 knots [surface], 7.7 knots [submerged]
Main Armament:	Six 533mm (21in) torpedo tubes, one 99mm (9.4in) gun
Launched:	1941

Bronzo, a Series 600 boat of the *Acciaio* class, operating in the Mediterranean, was captured by four British minesweepers on 12 July 1943 after surfacing off Syracuse. After a spell as the British submarine *P714,* it was handed over to the Free French in January 1944, and renamed *Narval*. It was broken up in 1948.

Seraph

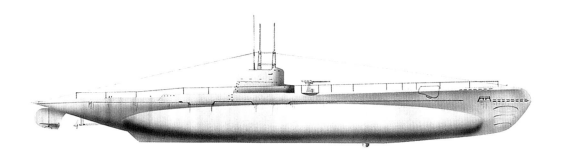

SPECIFICATIONS	
Type:	British submarine
Displacement:	886 tonnes (872 tons) (surface), 1,005 tonnes (990 tons) (submerged)
Dimensions:	66.1m x 7.2m x 3.4m (216ft 10in x 23ft 8in x 11ft 2in)
Machinery:	Twin screws, diesels (surface), electric motors (submerged)
Top speed:	14.7 knots [surface], 9 knots [submerged]
Main armament:	One 76mm (3in) gun, six 533mm (21in) torpedo tubes
Launched:	1941

Seraph was a medium-range boat with a diving depth of 95m (310ft). The diesel engines developed 1900hp, and surfaced range at 10 knots was 11,400km (6000 miles). The class, numbering 63 units, gave excellent service both during and after World War II. *Seraph* was broken up in 1965.

Drum

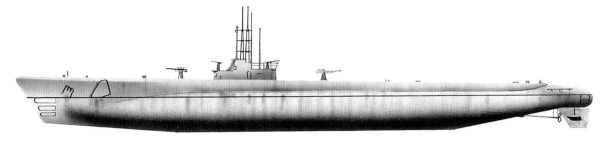

SPECIFICATIONS	
Type:	US submarine
Displacement:	1854 tonnes (1825 tons) [surface], 2448 tonnes (2410 tons) [submerged]
Dimensions:	95m x 8.3m x 4.6m (311ft 9in x 27ft 3in x 15ft 3in)
Machinery:	Twin screws, diesel [surface], electric motors [submerged]
Top speed:	20 knots [surface], 10 knots [submerged]
Main armament:	Ten 533mm (21in) torpedo tubes, one 76mm (3in) gun
Launched:	May 1941

Construction of the *Drum* class was the largest warship project undertaken by the United States: over 300 were built. Double hulled, ocean-going, they were excellent submarines with good seakeeping qualities and range. *Drum* was preserved on decommissioning, and has been a museum ship since 1968.

World War II Submarines: Part 2

A relatively small number of German U-boats operating in 'wolf packs' sank 712 ships between June 1942 and June 1943. But from the end of 1942, U-boats were being sunk at an increasing rate, with aircraft accounting for half of the losses. In May 1943, 41 U-boats were destroyed. They had lost the Battle of the Atlantic.

Grouper

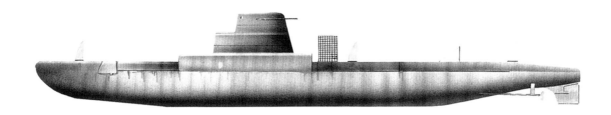

In 1951, *Grouper* was converted into one of the first hunter/killer submarines. The concept required a boat to be quiet and carry long-range sonar with high bearing accuracy. It could lie off enemy bases, or in narrow straits, and intercept hostile submarines. In 1958, it became a test vessel, and was scrapped in 1970.

SPECIFICATIONS

Type:	US submarine
Displacement:	1845 tonnes (1816 tons) [surface], 2463 tonnes (2425 tons) [submerged]
Dimensions:	94.8m x 8.2m x 4.5m (311ft 3in x 27ft 15ft)
Machinery:	Twin screws, diesel engines [surface], electric motors [submerged]
Top speed:	20.25 knots [surface], 10 knots [submerged]
Launched:	October 1941

Enrico Tazzoli

Enrico Tazzoli began as the US submarine *Barb,* completed in 1943 as part of the World War II *Gato* class. It was transferred to the Italian Navy in 1955 after it was fitted with the Guppy snorkel, including a modified structure and 'fairwater' for better underwater performance. Range was 19,311km (12,000 miles) at 10 knots.

SPECIFICATIONS

Type:	Italian submarine
Displacement:	1845 tonnes (1816 tons) [surface], 2463 tonnes (2425 tons) [submerged]
Dimensions:	94m x 8.2m x 5m (311ft 3in x 27ft x 17ft)
Machinery:	Twin screws, diesel engines [surface], electric motors [submerged]
Top speed:	20 knots [surface], 10 knots [submerged]
Main armament:	Ten 533mm (21in) torpedo tubes
Launched:	April 1942

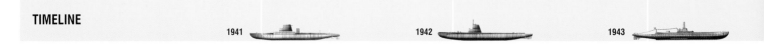

TIMELINE   1941   1942   1943

C1

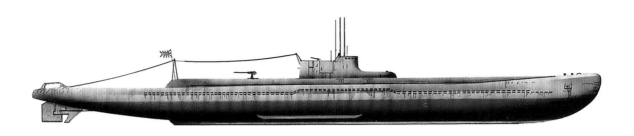

This large group of submarines was laid down from the early 1940s. Most were patrol boats; some became supply transports for beleaguered Japanese forces on the Pacific islands. Later, some were converted to carry four of the small *Kaiten*, or suicide boats, on the rear hull casing just aft of the conning tower.

SPECIFICATIONS	
Type:	Japanese submarine
Displacement:	2605 tonnes (2564 tons) [surface], 3702 (3761 tons) [submerged]
Dimensions:	108.6m x 9m x 5m (256ft 3in x 29ft 5in x 16ft 4in)
Machinery:	Twin screws, diesel engines [surface], electric motors [submerged]
Top speed:	17.7 knots [surface], 6 knots [submerged]
Main armament:	Six 533mm (21in) torpedo tubes, one 140mm (5.5in) gun
Launched:	1943

CB12

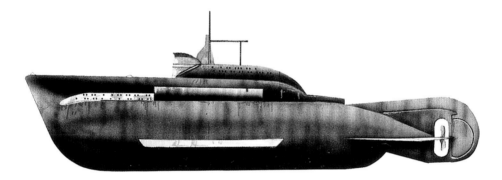

The CB programme of miniature submarines was begun in 1941. Its operating range was 2660km (1400 miles) at 5 knots on diesel engines, 95km (50 miles) at 3 knots submerged. Maximum diving depth was 55m (180ft). Complement was one officer and three crew. In all, 22 units were built, by Caproni Taliedo of Milan.

SPECIFICATIONS	
Type:	Italian midget submarine
Displacement:	25 tonnes (24.9 tons) [surface], 36 tonnes (35.9 tons) [submerged]
Dimensions:	15m x 3m x 2m (49ft 3in x 9ft 10in x 6ft 9in)
Machinery:	Single screw, diesel engine [surface], electric motor [submerged]
Top speed:	7.5knots [surface], 6.6 knots [submerged]
Main armament:	Two 450mm (17.7in) torpedoes in exterior cages
Launched:	August 1943

Diablo

A double-hulled ocean-going submarine, *Diablo* was a development of the *Gato* class, more strongly built and with improvements to the internal layout, which increased the displacement by about 40.5 tonnes (40 tons). Transferred to Pakistan, it was renamed *Ghazi* in 1964. It was sunk in the 1971 war with India.

SPECIFICATIONS	
Type:	US submarine
Displacement:	1890 tonnes (1860 tons) [surface], 2467 tonnes (2420 tons) [submerged]
Dimensions:	95m x 8.3m x 4.6m (311ft 9in x 27ft 3in x 15ft 3in)
Machinery:	Twin screws, diesel engines [surface], electric motors [submerged]
Top speed:	20 knots [surface], 10 knots [submerged]
Main armament:	Ten 533mm (21in) torpedo tubes, two 150mm (5.9in) guns
Launched:	1944

1944

U-47

U-47 on 13–14 October 1939 penetrated Scapa Flow and sank the 27,940-tonne (27,500-ton) battleship *Royal Oak.* It had already sunk three small merchant ships, and accounted for 27 more before it disappeared in the North Atlantic on 7 March 1941, perhaps sunk by the corvettes HMS *Arbutus* and *Camellia.*

U-47

U-47 was an early Type VIIB submarine, which in 1939 sunk the battleship HMS *Royal Oak* inside Scapa Flow. It was one of the most daring submarine attacks of the war and a bitter blow to British prestige.

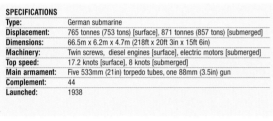

SPECIFICATIONS

Type:	German submarine
Displacement:	765 tonnes (753 tons) [surface], 871 tonnes (857 tons) [submerged]
Dimensions:	66.5m x 6.2m x 4.7m (218ft x 20ft 3in x 15ft 6in)
Machinery:	Twin screws, diesel engines [surface], electric motors [submerged]
Top speed:	17.2 knots [surface], 8 knots [submerged]
Main armament:	Five 533mm (21in) torpedo tubes, one 88mm (3.5in) gun
Complement:	44
Launched:	1938

RANGE
A good range was important: U-47 could run for 15,660km (8700 nautical miles) at a surface speed of 10 knots.

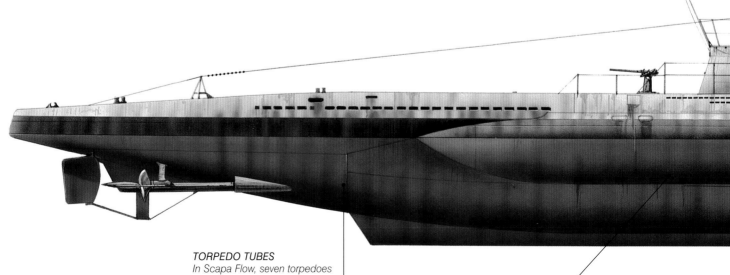

TORPEDO TUBES
In Scapa Flow, seven torpedoes were fired, from both bow and stern tubes; four struck, three in the captain's words 'went to blazes.'

MACHINERY
The two diesel engines developed 1MW (1400 shp) on the surface, and two electric motors generated 280kW (375 shp) underwater.

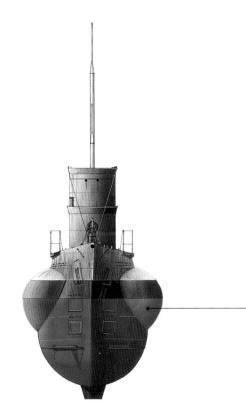

HULL
Compared with Type VIIA, the VIIB boats carried an extra 33 tonnes (32.5 tons) of oil in external saddle tanks, increasing their range.

TORPEDOES
Despite the successes, there were problems with German torpedoes, centred on faults in the magnetic pistol and a tendency to run too deep.

GUNS
The 88mm (3.5in) gun was removed from some boats in favour of an AA gun, as the air threat to submarines increased.

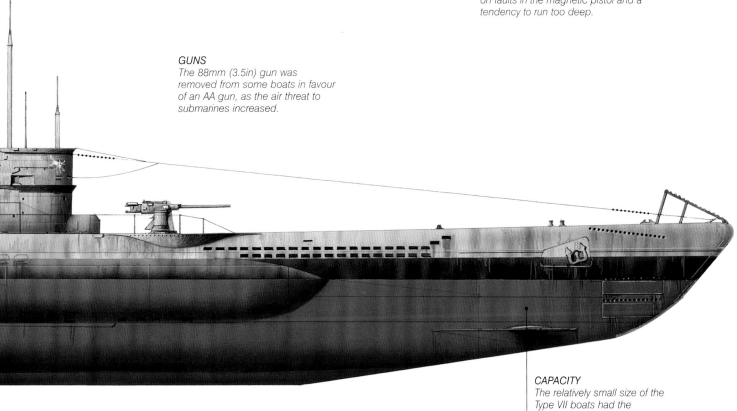

CAPACITY
The relatively small size of the Type VII boats had the disadvantage of limiting torpedo capacity to 14. Gunnery was used whenever possible.

World War II Submarines: Part 3

Submarine warfare racked up daunting statistics. US submarines sank about 1300 Japanese ships (5.5 million tonnes), and U-boats sank 14.6 million tonnes of Allied shipping, mostly in the Atlantic. Casualty rates among submariners were the highest of all services: among US crews, it was 22 per cent, and among German crews, 63 per cent.

I201

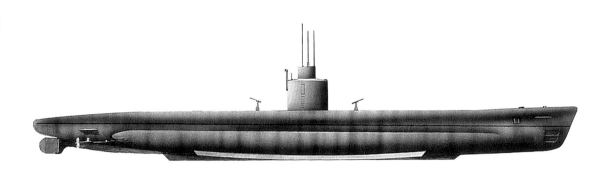

I201, with a smooth all-welded hull, was twice as fast underwater as US contemporaries – but heavy electricity consumption was the price. Surface range was 15,200km (8000 miles) at 11 knots, but submerged it was only 256km (135 miles) at 3 knots. In August 1945, it surrendered to US forces.

SPECIFICATIONS

Type:	Japanese submarine
Displacement:	1311 tonnes (1291 tons) [surface], 1473 tonnes (1450 tons) [submerged]
Dimensions:	79m x 5.8m x 5.4m (259ft 2in x 19ft x 17ft 9in)
Machinery:	Twin screws, diesel engines [surface], electric motors [submerged]
Top speed:	15.7 knots [surface], 19 knots [submerged]
Main armament:	Four 533mm (21in) torpedo tubes
Complement:	31
Launched:	1944

I351

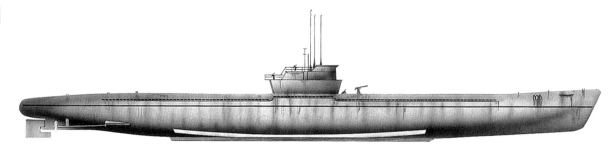

I351 was intended to provide support and supplies to seaplanes in forward areas where shore facilities and surface ships were not available. Up to 300 tonnes of aviation fuel could be carried. Two units were built. *I351* was sunk by the US submarine *Bluefish* on 14 July 1945, after only six months in service.

SPECIFICATIONS

Type:	Japanese submarine
Displacement:	3568 tonnes (3512 tons) [surface], 4358 tonnes (4290 tons) [submerged]
Dimensions:	110m x 10.2m x 6m (361ft x 33ft 6in x 20ft)
Machinery:	Twin screws, diesel engines [surface], electric motors [submerged]
Top speed:	15.7 knots [surface], 6.3 knots [submerged]
Main armament:	Four 533mm (21in) torpedo tubes
Complement:	90
Launched:	1944

TIMELINE

1944

I400

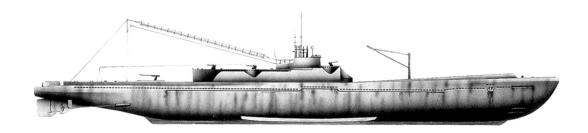

Three floatplanes were carried and could be launched from this class (of two), the largest ever diesel-electric submarines. *I400* would surface, the machines would be warmed up in the hangar, then rolled forward, their wings unfolded, and launched off a catapult rail. All aircraft could be airborne in 45 minutes.

SPECIFICATIONS

Type:	Japanese submarine
Displacement:	5316 tonnes (5233 tons) [surface], 6665 tonnes (6560 tons) [submerged]
Dimensions:	122m x 12m x 7m (400ft 3in x 39ft 4in x 23ft)
Machinery:	Twin screws, diesel engines [surface], electric motors [submerged]
Top speed:	18.7 knots [surface], 6.5 knots [submerged]
Main armament:	Eight 533mm (21in) torpedo tubes, one 140mm (5.5in) gun
Aircraft:	Three M6A1 Seiran floatplanes, plus components for a fourth
Launched:	1944

U-2501

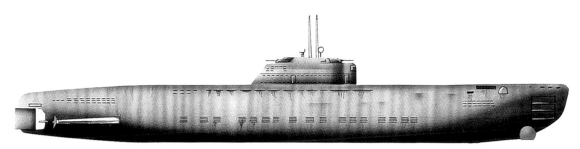

The Type XXI was double-hulled, with a high submerged speed, and ran silently at 3.5 knots. The outer hull was of light plating. The inner hull was of 28–37mm (1–1.5in) carbon steel plating. Batteries let it run submerged for three days on a single charge. By 1945, 55 had entered service. *U-2501* was scuttled in 1945.

SPECIFICATIONS

Type:	German submarine
Displacement:	1647 tonnes (1621 tons) [surface], 2100 tonnes (2067 tons) [submerged]
Dimensions:	77m x 8m x 6.2m (251ft 8in x 26ft 3in x 20ft 4in)
Machinery:	Twin screws, diesel engines [surface], electric motors [submerged]
Top speed:	15.5 knots [surface], 10 knots [submerged]
Main armament:	Six 533mm (21in) torpedo tubes, four 30mm (1.18in) guns
Launched:	1944

Entemedor

The *Gato* class had been planned in December 1941, and 54 submarines were built in an accelerated programme. Fuel tanks, with up to 480 tonnes (472 tons), were sited in the central double hull section. Maximum diving depth was 95m (312ft). *Entemedor* was transferred to Turkey in 1973.

SPECIFICATIONS

Type:	US submarine
Displacement:	1854 tonnes (1825 tons) [surface], 2458 tonnes (2420 tons) [submerged]
Dimensions:	95m x 8.3m x 4.6m (311ft 9in x 27ft 3in x 15ft 3in)
Machinery:	Twin screws, diesel engines [surface], electric motors [submerged]
Top speed:	20.2 knots [surface], 8.7 knots [submerged]
Main armament:	Ten 533mm (21in) torpedo tubes
Launched:	December 1944

Passenger Liners

Questions of national rivalry and prestige, as well as strict commerce, entered into the construction of large transatlantic liners. France's *Normandie*, Italy's *Rex* and *Conte di Savoia*, and Britain's *Queen Mary* all entered service in the 1930s, each claiming ways in which they were the best, or biggest, ships of the day.

Empress of Britain

The Canadian Pacific's largest passenger vessel, it could carry a total of 1195 passengers. In 1939, it was taken over as a troop transport. Attacked by a German bomber off the Irish coast in October 1940, it was taken in tow by the Polish destroyer *Burza*. However, on 28 October it was sunk by submarine *U-32*.

SPECIFICATIONS

Type:	Canadian liner
Displacement:	43,025 tonnes (42,348 tons)
Dimensions:	231.8m x 29.7m (760ft 6in x 97ft 5in)
Machinery:	Quadruple screws, geared turbines
Top speed:	25.5 knots
Route:	Canada–Europe
Launched:	June 1930

Georges Philippar

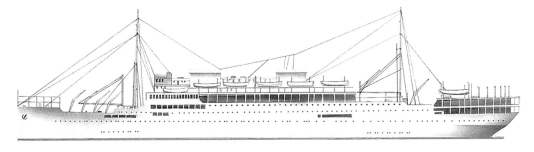

The motor ship *Georges Philippar* was short-lived. Built for Messageries Maritimes of Marseille, it made its maiden voyage to the Far East in February 1932. During the return trip, an electrical fire broke out, and the vessel sank 233km (145 miles) north-east of Cape Guardafui. Fifty-four passengers died.

SPECIFICATIONS

Type:	French liner
Displacement:	17,819 tonnes (17,539 tons)
Dimensions:	172.7m x 20.8m (19ft 10in x 68ft 3in)
Machinery:	Twin screws, diesel engines
Top speed:	17 knots
Launched:	November 1930

TIMELINE

1930   1931

Rex

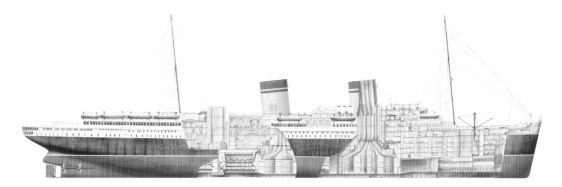

Italy's largest passenger liner, *Rex* was an emblem of national prestige, intended to capture the Atlantic Blue Riband, which it did until the arrival of the French *Normandie* in 1935. Laid up from the spring of 1940, it was sunk by RAF planes at Trieste in September 1944 and was broken up in 1947.

SPECIFICATIONS	
Type:	Italian liner
Displacement:	49,278 tonnes (48,502 tons)
Dimensions:	248.3m x 30m (814ft 8in x 96ft 2in)
Machinery:	Quadruple screws, geared turbines
Top speed:	29.5 knots
Route:	Genoa–New York
Launched:	October 1931

Conte di Savoia

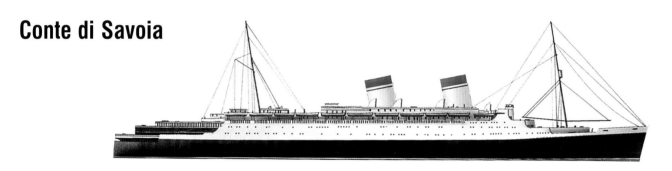

A stylish and elegantly fitted liner, built at Trieste, carrying up to 2060 passengers, *Conte di Savoia* was equipped with a gyro-driven stabilizer, the first to be used on so large a vessel. Laid up near Venice in 1939, it was sunk by Allied bombing in September 1943. Raised in 1946, it was scrapped in 1950.

SPECIFICATIONS	
Type:	Italian liner
Displacement:	51,879 tonnes (51,062 tons)
Dimensions:	268m x 29.25m x 8.55m (879ft x 96ft x 28ft)
Machinery:	Quadruple screws, geared turbines
Service speed:	27 knots
Route:	Genoa–New York
Complement:	750
Launched:	October 1931

Nieuw Amsterdam

Berthed at New York when World War II broke out, *Nieuw Amsterdam* served as a troopship when the United States entered the war, and covered over 800,000km (500,000 miles) in this role. Refitted after the war, it resumed regular transatlantic service. From 1971, it was a cruise liner, until scrapped in 1974.

SPECIFICATIONS	
Type:	Dutch liner
Displacement:	36,867.6 tonnes (36,287 tons)
Dimensions:	321.2 x 26.9m (758ft 6in x 88ft 4in)
Machinery:	Twin screws, geared turbines
Top speed:	20.5 knots
Route:	Rotterdam–New York
Launched:	1937

1937

Queen Mary

Built to be the biggest and fastest, *Queen Mary* held the Blue Riband until 1952. It carried 2739 passengers, but as a wartime troopship some 15,000 servicemen could be accommodated. After 1001 Atlantic crossings in mercantile service, it was sold in 1967 to Long Beach, California, where it is moored as a museum-cum-hotel.

VENTILATION
Ventilation funnels and fans brought cooler air down to the boiler spaces. Ships of this period were not air-conditioned.

SPEED
In wartime, Queen Mary's high speed (31.7 knots on occasion) was its prime defence against attack from submarines or surface ships.

MACHINERY
24 Yarrow boilers served four sets of Parsons single-reduction geared steam turbines. Maximum power output was 119.3MW (160,000 shaft horse-power).

ACCOMMODATION
On one troopship voyage, 16,683 soldiers and crew were packed in: the maximum passenger number ever carried.

Queen Mary

The ship was named after Queen Mary, the consort of King George V. Until the launch the name it was to be given was kept a closely guarded secret. Legend has it that Cunard intended to name the ship 'Victoria', in keeping with company tradition of giving its ships names ending in 'ia'.

SPECIFICATIONS

Type:	British liner
Displacement:	82,537 tonnes (81,237 tons)
Dimensions:	310.75 x 36.15m (1019ft 6in x 118ft 7in)
Machinery:	Quadruple screws, geared turbines
Top speed:	29 knots
Route:	Southampton–New York
Launched:	1934

FIRST-CLASS AREA
An indoor swimmming pool, cinema, cocktail lounge, library and drawing room, squash and deck tennis courts were among the amenities.

PASSENGERS
As in virtually all pre-war liners there were three classes; First: 776, Tourist: 784, Third: 579. The crew numbered 1071.

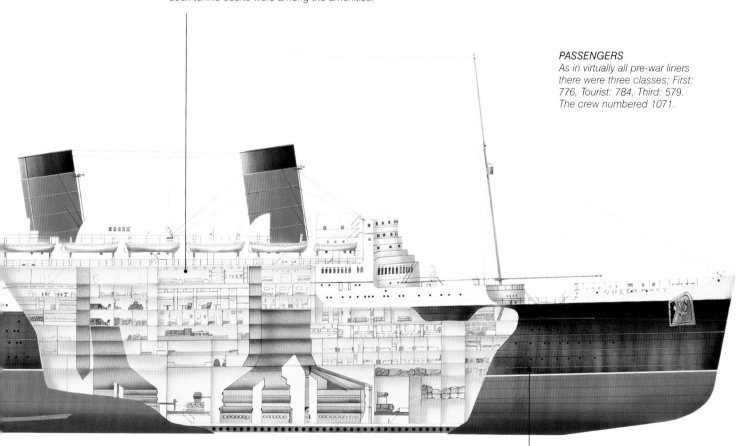

HULL
Queen Mary was the second 1000ft (300m)-plus liner, after the French Normandie (1935) which was ten feet (3m) longer.

Passenger–Cargo Liners

The depression years of the 1930s saw a reduction in merchant shipbuilding generally. Construction of some ships was delayed, and many vessels were laid up for a few years. In new ships, the accent was on economy and efficiency. Oil fuel replaced coal and crew numbers were reduced.

Empire Windrush

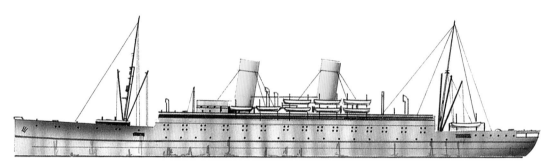

Launched as the German liner *Monte Rosa*, this was a troopship in 1942, then a repair workshop for the battleship *Tirpitz*. Damaged in 1944 by mines, it became a hospital ship. After 1945, as the British liner *Empire Windrush*, it brought the first immigrants from the West Indies. In March 1954, the ship caught fire and sank.

SPECIFICATIONS	
Type:	British liner
Displacement:	14,104 tonnes (13,882 tons)
Dimensions:	160m x 20m (524ft x 66ft)
Machinery:	Twin screws, geared diesels
Top speed:	14.5 knots
Launched:	1930

Clan Macalister

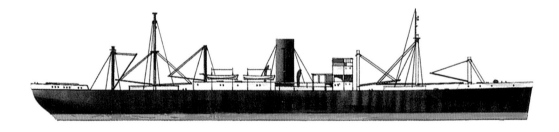

This general-purpose cargo liner was built for the Clan Line. In May 1940, it was commandeered to carry eight landing craft to Dunkirk to assist in the evacuation of British troops. Having unloaded the landing craft, it was attacked by German aircraft and caught fire. Efforts to control the blaze failed, and it was abandoned.

SPECIFICATIONS	
Type:	British cargo vessel
Displacement:	6896 tonnes (6787 tons)
Dimensions:	138m x 19m (453ft 8in x 62ft 3in)
Machinery:	Twin screws, triple expansion engines
Routes:	Britain–Africa and Far East
Launched:	1930

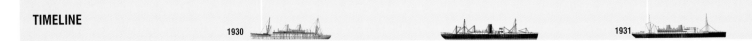

TIMELINE

1930

1931

Europa

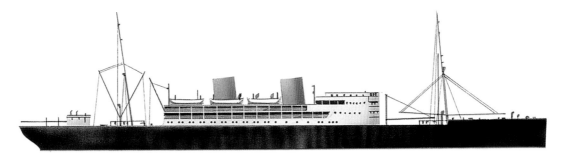

SPECIFICATIONS	
Type:	Danish liner
Displacement:	10,387 tonnes (10,224 tons)
Dimensions:	147.6m x 19m (484ft 3in x 62ft 4in)
Machinery:	Single screw, diesel engines
Top speed:	17.2 knots
Cargo:	General plus 64 passengers
Routes:	North Atlantic
Launched:	1931

The motor ship was popular with smaller shipping companies, like Denmark's East Asiatic Co., being more economical to run, and needing fewer engine-room staff. When Denmark was occupied in 1940, *Europa* transferred to the British flag. In May 1941, it was burnt out at Liverpool docks after a German air raid.

Carthage

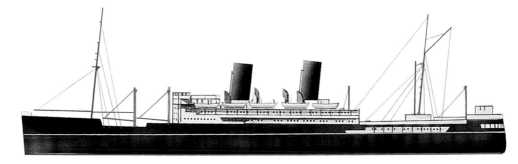

SPECIFICATIONS	
Type:	British liner
Displacement:	14,533 tonnes (14,304 tons)
Dimensions:	165m x 22m (540ft x 71ft)
Machinery:	Twin screws, geared turbines
Top speed:	19.5 knots
Route:	London–Far East ports
Cargo:	General cargo
Launched:	August 1931

As *Canton*, *Carthage* ran between London and Hong Kong from 1931, carrying passengers and mixed cargo. In 1940, it was refitted as an armed merchant cruiser with anti-aircraft weapons. In 1943, it was a troop transport. Renovated in 1947–48, it returned to commercial service. It was broken up in Japan in 1961.

Derbyshire

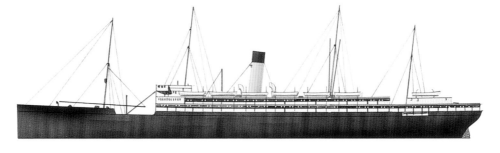

SPECIFICATIONS	
Type:	British liner
Displacement:	11,836 tonnes (11,650 tons)
Dimensions:	153m x 20m (502ft x 66ft 4in)
Machinery:	Twin screws, diesel engines
Top speed:	15 knots
Cargo:	General merchandise, rice, teak
Route:	London–Indian ports and Rangoon
Launched:	June 1935

Derbyshire was a one-class liner; its 291 passengers were mostly government employees travelling to and from British stations overseas. From 1939 to 1942, it served as an armed merchant cruiser, and then became a troopship. Refitted in 1946, it was a passenger/cargo liner 1948 to 1963, and was then sold for scrap.

 1935

SHIP TYPES 1950–1999

A significant development of this era was the digital computer and subsequent advances in electronic technology, which were seized on in ship design, construction and operation.

Huge merchant ships with automated navigational, mechanical and cargo-handling systems could be operated by a handful of crewmen. Satellite tracking and GPS locating were incorporated into navigation. In other developments, the passenger liner, made redundant by wide-bodied jet aircraft, evolved into the cruise ship. In naval operations, the nuclear reactor and the ballistic missile prompted a fundamental change of strategy, based on the nuclear submarine.

Left: *Nautilus* was the world's first nuclear-powered submarine. Early trials saw records broken including the first submerged transit across the North Pole.

Aircraft Carriers – USA

Among surface warships, the aircraft carrier took on the role of biggest and most dominant. The US Navy had many more and much larger carriers than any other fleet. Its *Kitty Hawk* and *Nimitz* classes, developed over decades, were vast ships with crew numbers running into thousands, and possessed of devastating firepower.

Forrestal

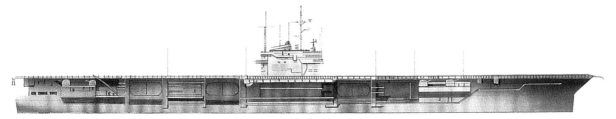

Forrestal and its three sisters were authorized in 1951. Large size was needed for combat jets, which needed more fuel than their piston-engined predecessors. *Forrestal* had space for 3.4 million litres (750,000 gallons) of aviation fuel. Decommissioned since 1993, it is to be sunk as a fishing reef.

SPECIFICATIONS

Type:	US aircraft carrier
Displacement:	80,516 tonnes (79,248 tons)
Dimensions:	309.4m x 73.2m x 11.3m (1015ft x 240ft x 37ft)
Machinery:	Quadruple screws, turbines
Top speed:	33 knots
Main armament:	Eight 127mm (5in) guns
Aircraft:	90
Complement:	2764 plus 1912 air wing
Launched:	December 1954

Enterprise

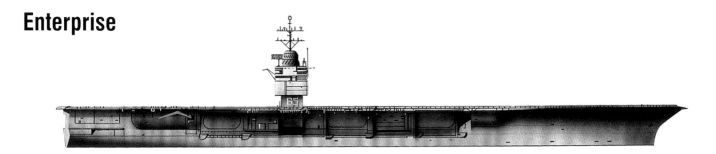

When completed in 1961, *Enterprise* was the largest ship in the world, and only the second nuclear-powered warship. Eight reactors gave it a range of 643,720km (400,000 miles) at 20 knots. It was re-fitted between 1979 and 1982, with a new island structure, and reconstructed in 1990–94 to serve until 2012–14.

SPECIFICATIONS

Type:	US aircraft carrier
Displacement:	91,033 tonnes (89,600 tons) full load
Dimensions:	335.2m x 76.8m x 10.9m (1100ft x 252ft x 36ft)
Machinery:	Quadruple screws, turbines, steam supplied by eight nuclear reactors
Top speed:	35 knots
Aircraft:	99
Complement:	5500 including air wing
Launched:	September 1960

TIMELINE

1954   1960   1964

America

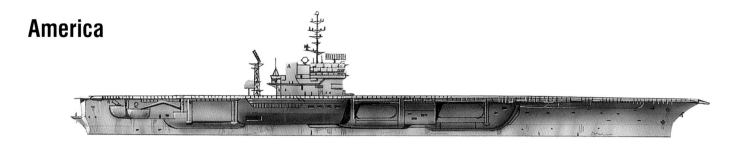

SPECIFICATIONS	
Type:	US aircraft carrier
Displacement:	81,090 tonnes (79,813 tons)
Dimensions:	324m x 77m x 10.7m (1063ft x 252ft 7in x 35ft)
Machinery:	Quadruple screws, geared turbines
Top speed:	33 knots
Main armament:	Three Mark 29 launchers for Sea Sparrow SAMs, three 20mm (0.79in) Phalanx CIWS
Aircraft:	82
Complement:	3306 excluding air wing
Launched:	1964

The *Kitty Hawk* class were the first carriers not to carry conventional guns. *America* was the first carrier equipped with an integrated Combat Information Centre (CIC). It took part in US engagements from Vietnam to Desert Storm (1991). Decommissioned in 1996, it was scuttled in 2005 after use as a target.

John F. Kennedy

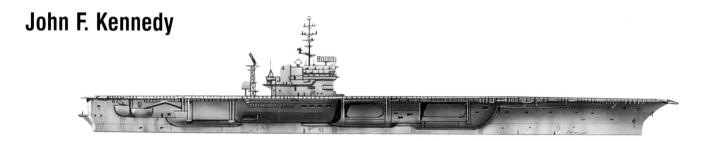

SPECIFICATIONS	
Type:	US carrier
Displacement:	82,240 tonnes (80,945 tons)
Dimensions:	320m x 76.7m x 11.4m (1052ft x 251ft 8in x 36ft)
Machinery:	Quadruple screws, geared turbines
Top speed:	33.6 knots
Main armament:	Three Sea Sparrow octuple launchers, three Mk15 Phalanx 20mm (0.79in) CIWS
Complement:	3306 plus 1379 air wing
Launched:	1967

Kennedy was the first to have an underwater protection system. Completed in May 1968, it was based in the North Atlantic and Mediterranean. In the 1980s, it was deployed off Lebanon and Libya, and off Iraq in 1991. It flew bombing missions against Al Qaeda targets in 2002. From 2007, it has been in the Reserve Fleet.

George Washington

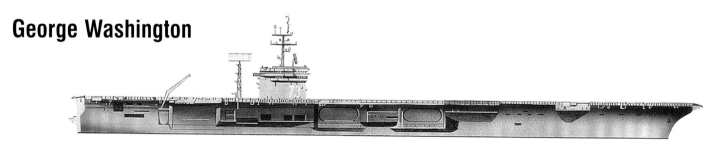

SPECIFICATIONS	
Type:	US aircraft carrier
Displacement:	92,950 tonnes (91,487 tons)
Dimensions:	332.9m x 40.8m x 11.3m (1092ft 2in x 133ft 10in x 37ft)
Machinery:	Quadruple screws, two water-cooled nuclear reactors, turbines
Top speed:	30 knots+
Main armament:	Four Vulcan 20mm (0.79in) guns plus missiles
Aircraft:	70+
Launched:	September 1989

A *Nimitz*-class supercarrier, *George Washington* carries damage-control systems, including armour 63mm (2.5in) thick over parts of the hull, plus box protection over the magazines and machinery spaces. Aviation equipment includes four lifts and four steam catapults. In 2010, it was operating from Yokosuka, Japan.

1967   1989

Invincible

Commissioned in 1980, *Invincible*'s 7° (later 12°) 'ski-jump' let its Sea Harriers take off at low, fuel-saving speed. It was deployed with the Falkland Islands Task Force in April–June 1982; in the Adriatic Sea during the Yugoslav wars in 1993–95; and off Iraq in 1988–99. *Invincible* was decommissioned in August 2005.

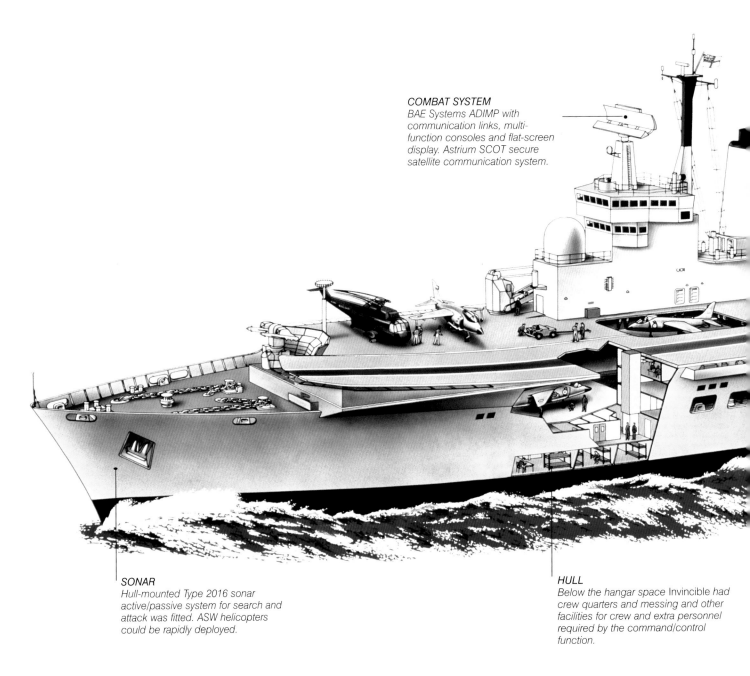

COMBAT SYSTEM
BAE Systems ADIMP with communication links, multi-function consoles and flat-screen display. Astrium SCOT secure satellite communication system.

SONAR
Hull-mounted Type 2016 sonar active/passive system for search and attack was fitted. ASW helicopters could be rapidly deployed.

HULL
Below the hangar space Invincible had crew quarters and messing and other facilities for crew and extra personnel required by the command/control function.

Invincible

The Royal Navy maintains that *Invincible* could be deployed should the need arise and that navy policy assumes that it is still an active aircraft carrier. But *Invincible* was stripped of some parts for sister ships, so bringing the ship to a state of operational readiness would require 18 months.

SPECIFICATIONS

Type:	British aircraft carrier
Displacement:	21,031 tonnes (20,700 tons)
Dimensions:	210m x 36m x 8.8m (689ft x 118ft 1in x 28ft 10in)
Machinery:	Twin screws, gas turbines
Top speed:	28 knots
Main armament:	Sea Dart anti-air and anti-missile missiles (removed c.1995), Goalkeeper CIWS
Aircraft:	Eight Harrier GR7/GR9, 11 Sea King helicopters
Complement:	726 plus 384 air wing
Launched:	1977

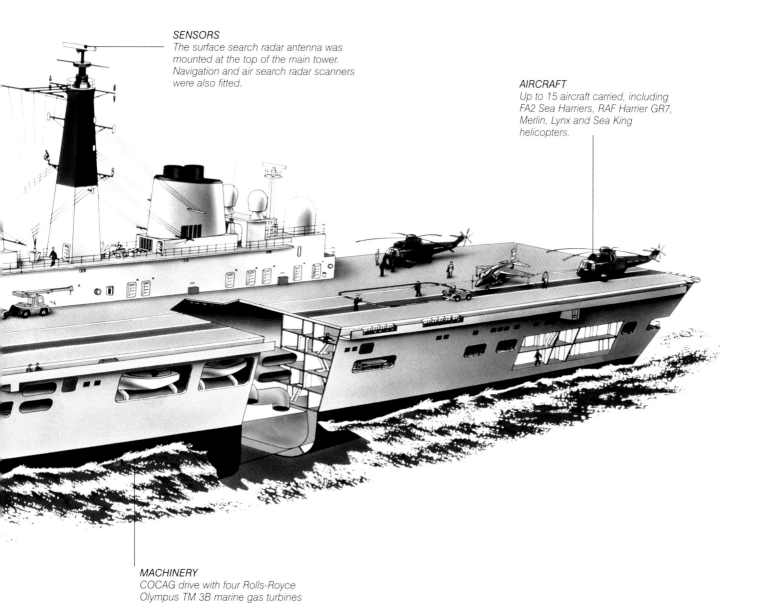

SENSORS
The surface search radar antenna was mounted at the top of the main tower. Navigation and air search radar scanners were also fitted.

AIRCRAFT
Up to 15 aircraft carried, including FA2 Sea Harriers, RAF Harrier GR7, Merlin, Lynx and Sea King helicopters.

MACHINERY
COCAG drive with four Rolls-Royce Olympus TM 3B marine gas turbines and eight Paxman Valenta diesel motors, producing 75MW (97,000hp).

Aircraft Carriers – Other Navies

Several carriers of World War II had unexpectedly long lives. Knowing the tactical importance of fixed-wing aircraft and helicopters, numerous countries sought to retain carriers or to acquire them second-hand. The introduction of angled flight decks and 'ski-jump' ramps enabled older ships to deploy modern aircraft.

Hermes

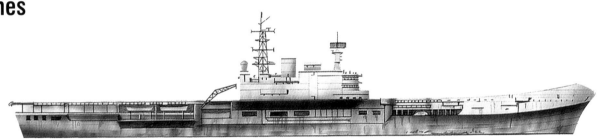

Plans for *Hermes* went back to 1943. After many design changes, the ship was finally completed in 1959. By 1979, it was handling the new Harrier vertical take-off jets. In 1982 it served as flagship in the Falklands War. In 1989, it was sold to India as *Viraat*, and was still operational in 2010.

SPECIFICATIONS

Type:	British aircraft carrier
Displacement:	25,290 tonnes (24,892 tons)
Dimensions:	224.6m x 30.4m x 8.2m (737ft x 100ft x 27ft)
Machinery:	Twin screws, turbines
Top speed:	29.5 knots
Main armament:	Thirty-two 40mm (1.6in) guns
Aircraft:	42
Launched:	February 1953

Clemenceau

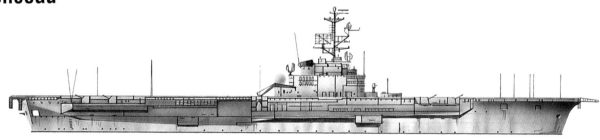

Clemenceau underwent modification during design and construction. It served in the Pacific, off Lebanon and in the 1991 Gulf War. Aircraft comprised 16 Super Etendards, three Etendard IVP, 10 F-8 Crusaders, seven Alize, plus helicopters. It was decommissioned in 2005 and sent for breaking in 2009.

SPECIFICATIONS

Type:	French aircraft carrier
Displacement:	33,304 tonnes (32,780 tons)
Dimensions:	257m x 46m x 9m (843ft 2in x 150ft x 28ft 3in)
Machinery:	Twin screws, geared turbines
Main armament:	Eight 100mm (3.9in) guns
Aircraft:	40
Launched:	December 1957

TIMELINE   1953   1957   1961

Vikrant

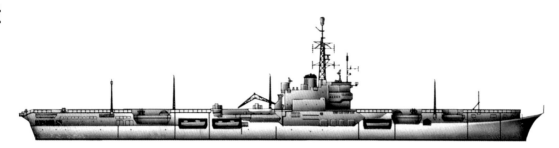

SPECIFICATIONS	
Type:	Indian aircraft carrier
Displacement:	19,812 tonnes (19,500 tons)
Dimensions:	213.4m x 39m x 7.3m (700ft x 128ftx 24ft)
Machinery:	Twin screws, geared turbines
Top speed:	23 knots
Main armament:	Fifteen 40mm (1.57in) cannon
Aircraft:	16
Complement:	1250
Launched:	1945 (modernized 1961)

Formerly the British light carrier *Hercules*, *Vikrant* was refitted in 1961 to carry Sea Hawk fighter-bombers. After the Indo-Pakistan war of 1971, it was refitted, including the provision of a ski-jump, in 1987–89, to carry Sea Harriers. It was withdrawn in 1996. India's new home-built carrier *Vikrant* is due for launch in 2010.

Veinticinco de Mayo

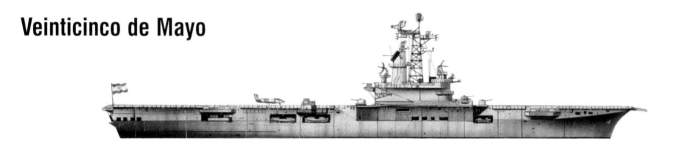

SPECIFICATIONS	
Type:	Argentinian aircraft carrier
Displacement:	20,214 tonnes (19,896 tons) full load
Dimensions:	211.3 x 36.9 x 7.6m (693ft 2in x 121ft x 25ft)
Machinery:	Twin screws, turbines
Top speed:	23 knots
Main armament:	Twelve 40mm (1.57in) cannon
Aircraft:	22
Complement:	1250
Launched:	1943 (modernized 1968)

First HMS *Venerable,* then the Dutch *Karel Doorman*, *Veinticinco de Mayo* was bought by Argentina in 1968. The flight deck was extended in 1979, and flew A-4Q Skyhawks, Super Etendards, S-2A Trackers and Sea King helicopters. It was engaged in the Falklands War of 1982. In 1997, it was broken up.

Kiev

SPECIFICATIONS	
Type:	Soviet aircraft carrier
Displacement:	38,608 tonnes (38,000 tons)
Dimensions:	273m x 47.2m x 8.2m (895ft 8in x 154ft 10in x 27ft)
Machinery:	Quadruple screws, turbines
Top speed:	32 knots
Main armament:	Four 76.2mm (3in) guns, plus missiles
Aircraft:	36
Launched:	December 1972

Completed in May 1975, *Kiev* was the first Russian aircraft carrier to be built with a full-length flight deck and a purpose-built hull. Apart from aircraft, it was armed with an array of missiles including the SS-N-12 Shaddock. Withdrawn in 1993, it has been an exhibit in a Chinese seaside theme park since 2004.

1968  1972

Giuseppe Garibaldi

Giuseppe Garibaldi has six decks with 13 watertight bulkheads. A 'ski-jump' launch ramp is mounted on the bows for vertical take-off and landing aircraft. This enables the aircraft to take off with a higher gross weight of fuel. It carries AV-8B Harrier jets or Agusta helicopters, or a combination of both, and has had several missile refits.

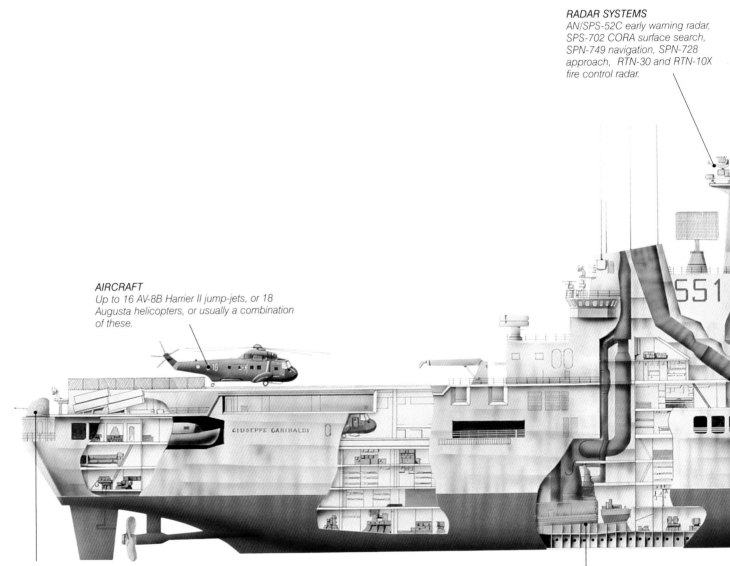

RADAR SYSTEMS
AN/SPS-52C early warning radar, SPS-702 CORA surface search, SPN-749 navigation, SPN-728 approach, RTN-30 and RTN-10X fire control radar.

AIRCRAFT
Up to 16 AV-8B Harrier II jump-jets, or 18 Augusta helicopters, or usually a combination of these.

GUNS
Three Selex NA 21 systems control three 40mm/70mm twin Oto Melara guns with an air target range of 4km (2.5 miles) and surface range of 12km (7.45 miles).

MACHINERY
COCAG drive with four Fiat-built General Electric LM2500 gas turbines and six diesel motors. Power output 60MW (81,000hp).

Giuseppe Garibaldi

SPECIFICATIONS	
Type:	Italian aircraft carrier
Displacement:	13,500 tonnes (13,370 tons)
Dimensions:	180m x 33.4m x 6.7m (590ft 6in x 109ft 6in x 22ft)
Machinery:	Quadruple screws, gas turbines
Top speed:	30 knots
Main armament:	Missile launchers, six torpedo tubes
Aircraft:	16 Harriers, or 18 helicopters
Complement:	550 plus 230 air wing
Launched:	1983

The WWII Peace Treaty banned Italy from having an aircraft carrier, which meant that at the time of launch *Giuseppe Garibaldi* did not receive its Harriers and was classed as an aircraft-carrying cruiser. The ban was eventually lifted and in 1989 the Italian Navy obtained fixed-wing aircraft to operate from the ship.

COUNTERMEASURES
SLQ-732 jamming system, SCLAR decoy launcher, SLAT anti-torpedo system, and SLQ-25 Nixie towed torpedo decoy.

SAM DEFENCE
Albatros eight-cell launchers are installed on the roof decks at the forward and stern end of the main island. 48 Aspide missiles with a range of 14km (8.6 miles) are carried.

FLIGHT DECK
The flight deck is 174m (570ft 10in) long and 30.5m (100ft) wide, and the forward 15m of the flight deck rises to a ski ramp of about 4°.

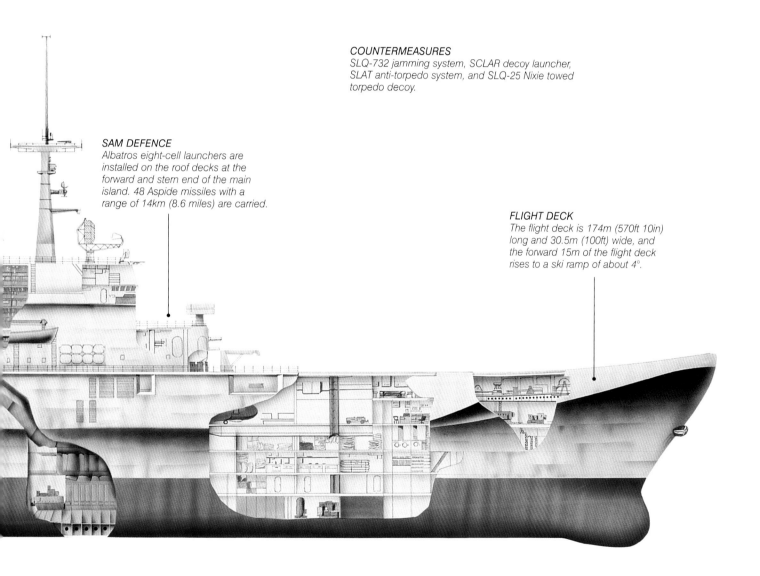

Helicopter Carriers & Small Aircraft Carriers

The vastly increased importance of the helicopter as an element in anti-submarine action has made it, in effect, an extendable arm of the modern warship. In addition, the helicopter carrier can transport, service and fuel a larger number of helicopters to take part in a various situations, from military invasion to humanitarian aid.

Iwo Jima

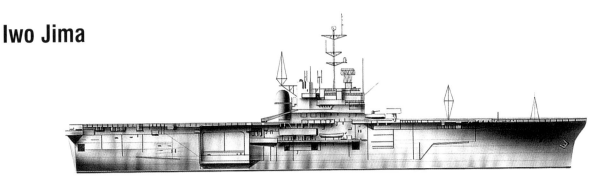

Iwo Jima was the first ship designed specifically to carry and operate helicopters, along with a Marine battalion of 2000 troops, plus artillery and support vehicles. In the 1970s, Sea Sparrow missile launchers were installed. A boiler explosion damaged the ship in 1990; it was stricken in 1993 and broken up in 1995.

SPECIFICATIONS	
Type:	US assault ship
Displacement:	18,330 tonnes (18,042 tons)
Dimensions:	183.6m x 25.7m x 8m (602ft 8in x 84ft x 26ft)
Machinery:	Single screw, turbines
Top speed:	23.5 knots
Main armament:	Four 76mm (3in) guns
Aircraft:	20 helicopters
Complement:	667, 2057 marines
Launched:	September 1960

Jeanne D'Arc

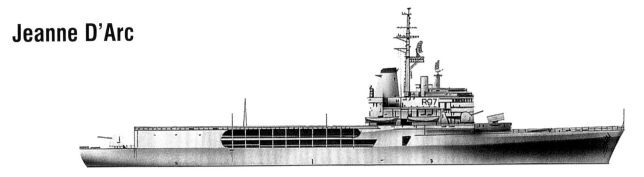

A multi-purpose cruiser, helicopter carrier and assault ship, *Jeanne D'Arc* could transport 700 men and eight large helicopters. In 1975, Exocet missiles were fitted, giving it a full anti-ship role. It also functioned as a training ship, providing facilities for up to 198 cadets at a time. It was struck from the list in 2009.

SPECIFICATIONS	
Type:	French helicopter carrier
Displacement:	13,208 tonnes (13,000 tons)
Dimensions:	180m x 25.9m x 6.2m (590ft 6in x 85ft x 20ft 4in)
Machinery:	Twin screws, turbines
Top speed:	26.5 knots
Main armament:	Four 100mm (3.9in) guns
Aircraft:	Eight
Complement:	627 including cadets
Launched:	September 1961

TIMELINE
1960
1961
1964

Moskva

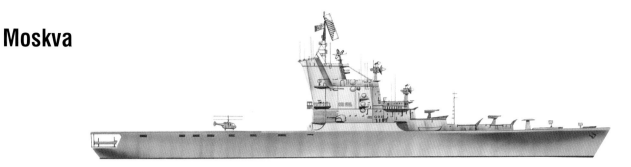

SPECIFICATIONS	
Type:	Soviet helicopter carrier
Displacement:	14,800 tonnes (14,567 tons)
Dimensions:	191m x 34m x 7.6m (626ft 8in x 111ft 6in x 25ft)
Machinery:	Twin screws, turbines
Top speed:	30 knots
Main armament:	One twin SUW-N-1 launcher, two twin SA-N-3 missile launchers
Aircraft:	18 helicopters
Complement:	850, including air wing
Launched:	1964

Moskva was the first helicopter carrier built for the Russian Navy, completed in 1967 to counteract the threat from the US nuclear-powered missile submarines that began to enter service in 1960. A central block dominated the vessel and housed the major weapons systems. *Moskva* was scrapped in the mid-1990s.

Vittorio Veneto

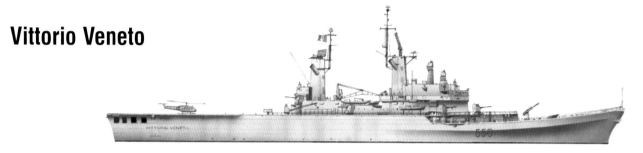

SPECIFICATIONS	
Type:	Italian helicopter cruiser
Displacement:	8991 tonnes (8850 tons)
Dimensions:	179.5m x 19.4m x 6m (589ft x 63ft 8in x 19ft 8in)
Machinery:	Twin screws, turbines
Top speed:	32 knots
Main armament:	Twelve 40mm (1.6in) guns, eight 76mm (3in) guns, four Teseo SAM launchers, one ASROC launcher
Aircraft:	Nine helicopters
Complement:	550
Launched:	February 1967

Vittorio Veneto was a purpose-built helicopter cruiser. A large central lift was set immediately aft of the superstructure, and two sets of fin stabilizers were fitted. Laid down in 1965, completed in 1969, the ship underwent a refit between 1981 and 1984. Withdrawn in 2003, it is intended to be a museum ship at Taranto.

Chakri Naruebet

SPECIFICATIONS	
Type:	Thai light carrier
Displacement:	11,480 tonnes (11,300 tons)
Dimensions:	182.5m x 30.5m x 6.15m (599ft 1in x 110ft 1in x 20ft 4in)
Machinery:	Twin screw, turbines and diesels
Top speed:	26 knots
Main armament:	Two launchers for Mistral SAM
Aircraft:	10
Complement:	455 plus 162 aircrew
Launched:	1996

Spanish-built, modelled on Spain's *Principe de Asturias*, Thailand's only carrier is also the world's smallest. *Chakri Naruebet* carries Harrier AV-8B VSTOL jump-jets and helicopters, the Harriers also being bought from Spain. The ship was in service in 2010, but its operational status is unclear and it rarely goes to sea.

1967

1996

Tarawa

Commissioned in 1976, equipped for air-land assault, *Tarawa* had a floodable well-deck for landing craft, and command and control facilities to undertake a flagship role. It was the first of a class of five, all built within the space of a few years in response to perceived threats during the Cold War.

Tarawa

USS *Tarawa* was the lead ship of the Navy's first class of amphibious assault ships able to incorporate the best design features and capabilities of several amphibious assault ships then in service.

MEDICAL FACILITIES
Tarawa's facilities included 300-bed hospital, four me operating rooms, and thre dental operating rooms.

SPECIFICATIONS

Type:	US assault ship
Displacement:	39,388 tonnes (38,761 tons) full load
Dimensions:	249.9m x 38.4m x 7.8m (820ft x 126ft x 25ft 9in)
Machinery:	Twin screws, geared turbines
Top speed:	24 knots
Main armament:	Two RAM launchers, two 127mm (5in) guns, two 20mm (0.79in) Phalanx CIWS
Aircraft:	Up to 35 helicopters, 8 AV-8B Harrier II
Complement:	892 plus 1093 troops
Launched:	1973

ARMAMENT
Four Mk38 Mod 1 25mm (0.98in) Bushmaster cannon, five M2HB 12.7mm (0.5in) calibre machine guns, two Mk15 Phalanx CIWS, and two Mk49 RAM launchers.

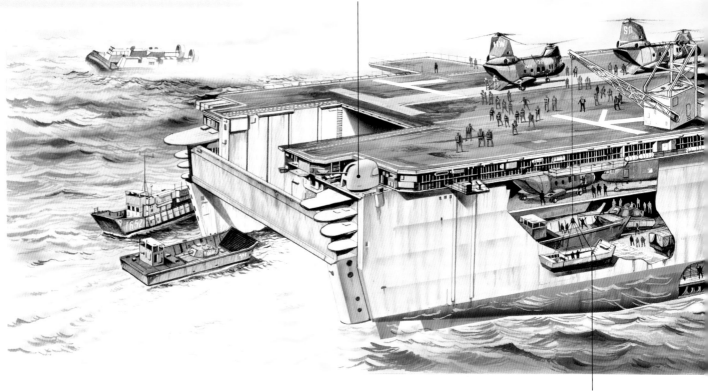

WELL DECK
Internal roadways enabled vehicles to be drive from the garage space to the landing craft loading points.

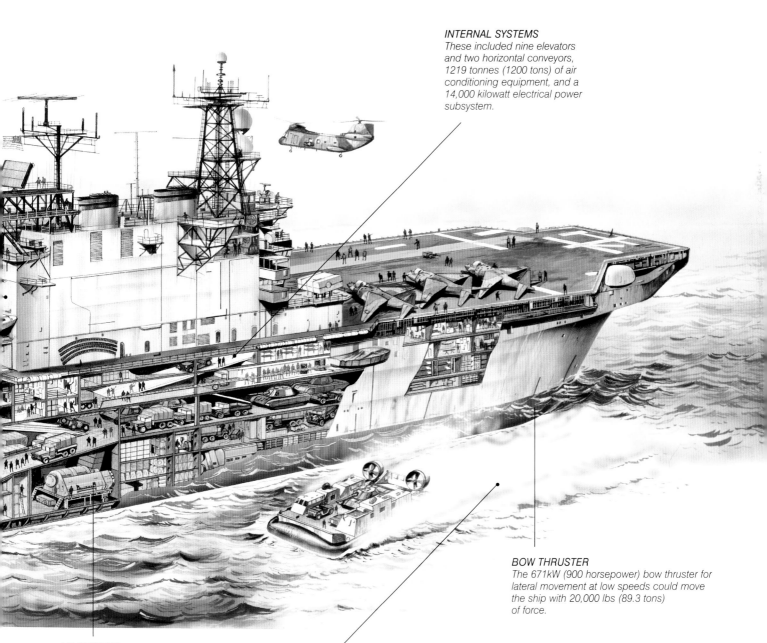

INTERNAL SYSTEMS
These included nine elevators and two horizontal conveyors, 1219 tonnes (1200 tons) of air conditioning equipment, and a 14,000 kilowatt electrical power subsystem.

BOW THRUSTER
The 671kW (900 horsepower) bow thruster for lateral movement at low speeds could move the ship with 20,000 lbs (89.3 tons) of force.

MACHINERY
The two boilers were the largest ever manufactured for the United States Navy, generating 406.4 tonnes (400 tons) of steam per hour, and developing 104,398kW (140,000hp).

BALLAST
Tarawa could ballast 12,192 tonnes (12,000 tons) of seawater for trimming the ship, while receiving and discharging landing craft from the well deck.

Assault Ships

The needs of amphibious warfare, mobilizing specialist equipment and specialized vehicles, and with sophisticated communications, have been answered by a new generation of landing ships and command ships. These have systems more developed than the LSTs and converted merchant ships of the 1940s and '50s.

Intrepid

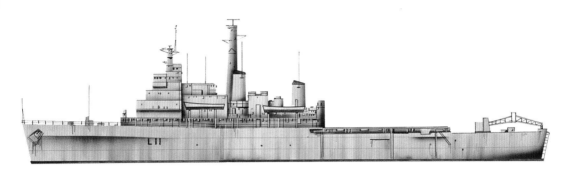

A 'landing platform dock', *Intrepid* could operate its own set of landing craft, and carry up to 700 troops. Above the landing dock were hangar and flight deck for six helicopters. In the Falklands War, the Argentinian surrender was signed on board *Intrepid*. Decommissioned in 1999, it was scrapped shortly after.

SPECIFICATIONS	
Type:	British assault ship (LPD)
Displacement:	12,313 tonnes (12,120 tons)
Dimensions:	158m x 24m x 6.2m (520ft x 80ft x 20ft 6in)
Machinery:	Twin screws, turbines
Top speed:	21 knots
Main armament:	Two 40mm (1.57in) guns, four Seacat anti-aircraft missile launchers
Complement:	566
Launched:	June 1964

Denver

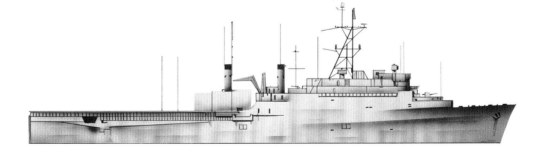

The 11 ships of this class, enlarged versions of the *Raleigh* group, have greater capacity for carrying troops and support vehicles. Assault and landing craft are held in the comprehensive docking facility forming the rear section of the vessel. In recent years, *Denver* has brought post-tsunami aid to Taiwan and Sumatra.

SPECIFICATIONS	
Type:	US command ship
Displacement:	9477 tonnes (9328 tons)
Dimensions:	174m x 30.5m x 7m (570ft 3in x 100ft x 23ft)
Machinery:	Twin screws, turbines
Top speed:	21 knots
Main armament:	Eight 76mm (3in) guns
Complement:	447 plus 840 marines
Launched:	January 1965

TIMELINE

1964  1965 1970

Mount Whitney

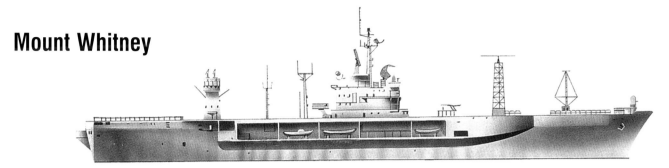

Modern warfare demands specialized command ships. *Mount Whitney*, flagship of the US Sixth Fleet from 2005, uses the same hull form and machinery as the Guam class, with flat open decks to allow maximum antenna placement. *Mount Whitney* has often been in 'hot-spot' situations, including the Black Sea in 2008.

SPECIFICATIONS	
Type:	US command ship
Displacement:	19,598 tonnes (19,290 tons)
Dimensions:	189m x 25m x 8.2m (620ft 5in x 82ft x 27ft)
Machinery:	Single screw, turbines
Top speed:	23 knots
Main armament:	Four 76mm (3in) guns, two eight-tube Sea Sparrow missile launchers
Complement:	700
Launched:	January 1970

Ivan Rogov

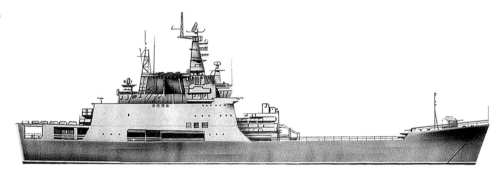

A long-range assault ship, carrying 550 troops, plus 40 tanks and other support vehicles, *Ivan Rogov* was fitted with a bow ramp, and a docking area 76m (250ft) long. The aft superstructure housed a helicopter. In 1979, it was transferred from the Black Sea to the Pacific Fleet, and stricken in 1996.

SPECIFICATIONS	
Type:	Soviet amphibious assault ship
Displacement:	13,208 tonnes (13,000 tons)
Dimensions:	158m x 24m x 8.2m (521ft 8in x 80ft 5in x 21ft 4in)
Machinery:	Twin screws, gas turbines
Top speed:	23 knots
Main armament:	Two 76mm (3in) guns, plus anti-aircraft missiles
Complement:	200
Launched:	1977

Whidbey Island

Whidbey Island's well deck accommodates four LCAC hovercraft or up to 21 smaller 61-tonne (60-ton) landing craft. The ship carries 450 troops, military vehicles and two assault and transport helicopters, and can fly Harrier jump jets. Assigned to Amphibious Group 2, it is currently deployed in the Persian Gulf.

SPECIFICATIONS	
Type:	US dock landing ship
Displacement:	15,977 tonnes (15,726 tons)
Dimensions:	186m x 25.6m x 6.3m (609ft x 84ft x 20ft 8in)
Machinery:	Twin screws, diesel engines
Top speed:	20+ knots
Main armament:	Two 20mm (0.79in) Vulcan guns
Complement:	340
Launched:	June 1983

1977  1983

Corvettes/Patrol Ships: Part 1

Seaward extension of international boundaries and increasing levels of smuggling provide a role for the patrol ship. Lightweight but effective missile launchers and rapid-fire automatic guns endow small ships with heavy fire-power. Reconnaissance, surveillance and interception are their prime tasks.

Shanghai

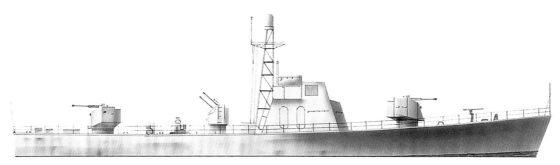

The Chinese Navy has many coastal patrol craft. The *Shanghai* vessels carry a relatively powerful armament of light weapons, plus depth charges and mines. Many have been exported to Asia, the Middle East and Africa, and others built under licence by European navies. Type 1 was in service until the early 1990s.

SPECIFICATIONS	
Type:	Chinese fast attack/patrol boat
Displacement:	137 tonnes (135 tons)
Dimensions:	38.8m x 5.4m x 1.7m (127ft 4in x 17ft 8in x 5ft 7in)
Machinery:	Quadruple screws, diesel engines
Top speed:	28.5 knots
Main armament:	Four 37mm (1.45in) guns, four 25.4mm (1in) cannon
Launched:	1962

Dardo

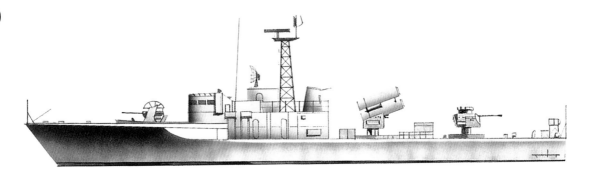

Dardo was one of four vessels whose functions could switch from minelayer (with an anti-aircraft gun and eight mines) to torpedo boat (with one 40mm/1.6in gun and 21 x 533mm/21in torpedoes). Conversion could be achieved in under 24 hours. Hybrid designs are not always successful, but these proved effective.

SPECIFICATIONS	
Type:	Italian motor gunboat
Displacement:	218 tonnes (215 tons)
Dimensions:	46m x 7m x 1.7m (150ft x 23ft 9in x 5ft 6in)
Machinery:	Twin screws, diesels and gas turbines
Top speed:	40+ knots
Main armament:	One 40mm (1.6in) gun, four 533mm (21in) torpedo tubes
Launched:	1964

TIMELINE

1962   1964   1966

Spica

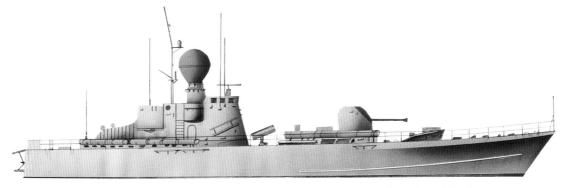

Spica was the first of a group of fast attack craft designed for Baltic waters. Bases are built into the rocky coastline, and can withstand most weapons except nuclear. The gas turbines develop 12,720hp, for rapid acceleration. The design was adopted by several other navies. *Spica* is now a museum ship in Stockholm.

SPECIFICATIONS	
Type:	Swedish fast attack torpedo craft
Displacement:	218 tonnes (215 tons)
Dimensions:	42.7m x 7m x2.6m (140ft x 23ft 4in x 8ft 6in)
Machinery:	Triple screws, gas turbines
Main armament:	One 57mm (2.24in) gun, six 533mm (21in) torpedo tubes
Launched:	1966

Nanuchka I

This class of light but heavily armed missile craft was known as the *Nanuchka I*. Some variants carried two 57mm guns and all had fire control radar and hull-mounted sonar systems. *Nanuchka II* types were built for India and III for Algeria and Libya. All Russian class members were decommissioned by the end of the 1990s.

SPECIFICATIONS	
Type:	Russian corvette
Displacement:	670.5 tonnes (660 tons)
Dimensions:	59.3m x 12.6m x 2.5m (194ft 7in x 41ft 4in x 7ft 11in)
Machinery:	Twin screws, diesels
Top speed:	32knots
Main armament:	Six SS-N-9 SSM, one SA-N-4 SAM launcher, one 76mm (3in) gun
Launched:	1969

D'Estienne d'Orves

D'Estienne d'Orves was one of a group of 20 frigates, small by contemporary standards, that followed on from the *Commandant Rivière* group, a class of larger ships. Designed for anti-submarine work in coastal waters, they can also operate at long range. This ship became the Turkish Navy's *Beykoz* in 1999.

SPECIFICATIONS	
Type:	French light frigate
Displacement:	1351 tonnes (1330 tons)
Dimensions:	80m x 10m x 3m (262ft 6in x 33ft 10in x 9ft 10in)
Machinery:	Twin screws, diesel engines
Top speed:	23.3 knots
Main armament:	Four Exocet launchers, one 100mm (3.9in) dual-purpose gun
Launched:	June 1973

1969   1975

Corvettes/Patrol Ships: Part 2

Longer-range patrol work needs a larger vessel than even a large motor boat, and several navies have responded with a modern version of the corvette. Once a slow light escort ship, it is now more likely to be a fast missile-bearing patrol craft with a full array of radar search and possibly sonar detection equipment.

Beskytteren

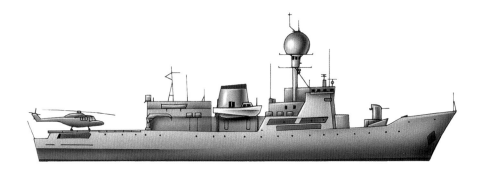

Designed for patrols in North Atlantic and Arctic waters, *Beskytteren* also has a fishery protection role. It is a smaller, modified version of the Danish *Hvidbjørnen* class frigate. Navigational radar and sonar equipment are fitted, and hangar and flight deck for a Lynx helicopter are incorporated, in a compact vessel.

SPECIFICATIONS

Type:	Danish patrol ship
Displacement:	2001.5 tonnes (1970 tons) full load
Dimensions:	74.4m x 12.5m x 4.5m (244ft x 41ft x 14ft 9in)
Machinery:	Single screw, three diesel motors
Top speed:	18 knots
Main armament:	One 76mm (3in) gun
Complement:	60
Launched:	1975

Fremantle

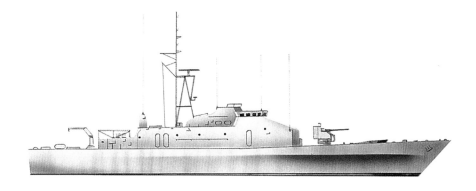

Lead ship of a class of 15, *Fremantle* was built in Lowestoft, England; the others were built at Cairns, Australia. They were faster than the *Attack*-class vessels that previously fulfilled the role of long-range coastal patrols to prevent smuggling and the landing of illegal immigrants. *Fremantle* served until 2006.

SPECIFICATIONS

Type:	Australian fast patrol boat
Displacement:	214 tonnes (211 tons)
Dimensions:	41.8m x 7.1m x 1.8m (137ft 2in x 23ft 4in x 6ft)
Machinery:	Triple screws, diesel engines
Top speed:	30 knots
Main armament:	One 40mm (1.6in) gun
Launched:	1979

TIMELINE

1975 　　1979 　　1980

Badr

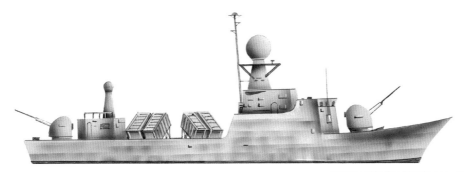

SPECIFICATIONS	
Type:	Egyptian fast patrol boat
Displacement:	355.6 tonnes (350 tons) full load
Dimensions:	52m x 7.6m x 2m (170ft 7in x 25ft x 6ft 7in)
Machinery:	Quadruple screws, four diesel engines
Top speed:	37 knots
Main armament:	Four SSM, one 76mm (3in) gun
Complement:	40
Launched:	1980

One of six boats packing a lot of hardware into a small hull, *Badr* has four OTO Melara/Matra Otomat Mk 1 anti-ship missiles, fired from box launchers mounted behind the deckhouse, and its weapons are directed by fire-control and target-tracking systems. The air-surface search radar dome is a conspicuous feature.

Kaszub

SPECIFICATIONS	
Type:	Polish corvette
Displacement:	1202 tonnes (1183 tons)
Dimensions:	82.3m x 10m x 2.8m (270ft 2in x 32ft 9in x 9ft 2in)
Machinery:	Quadruple screws, four diesel motors
Top speed:	28 knots
Main armament:	Two SA-N-5 SAM launchers, one 76mm (3in) gun, two 533mm (21in) torpedo tubes
Complement:	67
Launched:	1986

This Type 620 corvette, built in Poland, was to be the first of seven, but the collapse of the Warsaw Pact stopped development after *Kaszub*, which never received its intended Russian-made missile armament. Technical problems limit its usefulness, and it rarely puts to sea, though it is still on the active list in 2009.

Eilat

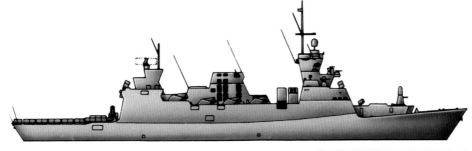

SPECIFICATIONS	
Type:	Israeli corvette
Displacement:	1295.4 tonnes (1275 tons) full load
Dimensions:	86.4m x 11.9m x 3.2m (283ft 6in x 39ft x 10ft 6in)
Machinery:	Twin screws, turbines plus turbo diesels
Top speed:	33 knots
Main armament:	Eight Harpoon SSM, eight Gabriel II SSM, one 76mm (3in gun)
Complement:	74
Launched:	1994

Designed on radar-dodging 'stealth' lines, *Eilat* is one of three well-armed small warships intended to lead Israel's large fleet of smaller attack craft. The forward gun position can be altered to mount different guns or a Phalanx CIWS system. A Dauphin helicopter is carried, and the vessel has sensor devices and radars.

1986    1994

USS Long Beach

Completed in September 1961, *Long Beach* was the United States' largest non-carrier surface warship built since 1945. Off Vietnam in 1968, it shot down two MiG fighters in the first successful SAM naval action. In the 1980s, Harpoon missiles and Phalanx CIWS were installed. *Long Beach* was withdrawn from service in 1994.

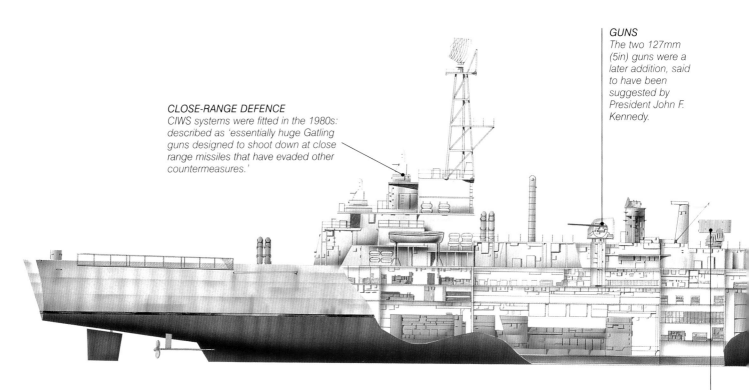

GUNS
The two 127mm (5in) guns were a later addition, said to have been suggested by President John F. Kennedy.

CLOSE-RANGE DEFENCE
CIWS systems were fitted in the 1980s: described as 'essentially huge Gatling guns designed to shoot down at close range missiles that have evaded other countermeasures.'

MACHINERY
Two CIW nuclear reactors powering two General Electric turbines. Total power output was 59,656kW (80,000shp). Range was effectively unlimited.

FINAL WEAPONS SUIT
Harpoon and BGM-109 Tomahawk missiles replaced the original Te and Talos systems. An 8 tube ASROC launcher v also fitted.

USS Long Beach

A deactivation ceremony took place on 2 July 1994 at Norfolk Naval Station. *Long Beach* was decommissioned on 1 May 1995, over 33 years after she had entered service. She is currently waiting to be recycled.

SPECIFICATIONS	
Type:	US missile cruiser
Displacement:	16,624 tonnes (16,602 tons)
Dimensions:	219.8m x 22.3m x 7.2m (721ft 3in x 73ft 4in x 23ft 9in)
Machinery:	Twin screws, two nuclear reactors driving geared turbines
Top speed:	30+ knots
Main armament:	Eight Harpoon SSM, two Terrier SSM systems, two 127mm (5in) guns, two 20mm (0.79in) Phalanx CIWS, one ASROC launcher, six 2324mm (12.75in) torpedo tubes
Complement:	1107
Launched:	1959

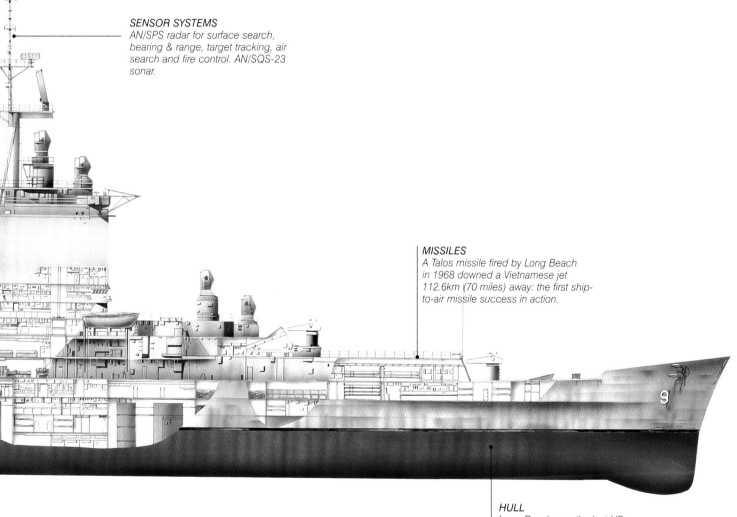

SENSOR SYSTEMS
AN/SPS radar for surface search, bearing & range, target tracking, air search and fire control. AN/SQS-23 sonar.

MISSILES
A Talos missile fired by Long Beach in 1968 downed a Vietnamese jet 112.6km (70 miles) away: the first ship-to-air missile success in action.

HULL
Long Beach was the last US cruiser to be built with a traditional cruiser-type hull, long and relatively narrow.

American Cruisers

While World War II marked the end of the battleship age, both the United States and Russia subsequently built many cruisers, though by the later 1950s the concept of the cruiser was obviously old-fashioned. New generation submarines, aircraft and missiles made them vulnerable. Smaller ships packed a more devastating punch.

Galveston

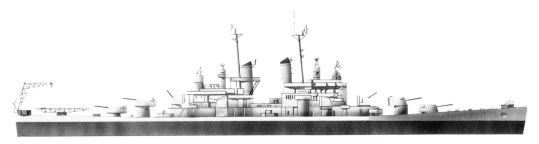

Galveston was one of six cruisers of World War II which were later refitted to carry Talos or Terrier missiles. The original guns were retained. The purpose was air defence and the ships were not expected to undertake deep sea missions. *Galveston* served until 1970, was stricken in 1973 and scrapped in 1975.

SPECIFICATIONS	
Type:	US cruiser
Displacement:	15,394 tonnes (15,152 tons)
Dimensions:	186m x 20m x 7.8m (610ft x 65ft 8in x 25ft 8in)
Machinery:	Quadruple screws, geared turbines
Top speed:	32 knots
Main armament:	Talos SAM system, six 152mm (6in) guns, six 127mm (5in) guns
Complement:	1382
Launched:	1945, converted 1958

Worden

Worden was one of nine vessels replacing cruisers from World War II. The layout incorporated masts and gantries for a complex radar system. The class was refitted in the late 1960s and again in the late 1980s. In 1991, *Worden* became an anti-aircraft command vessel. Stricken in 1993, it was sunk as a target in 2000.

SPECIFICATIONS	
Type:	US cruiser
Displacement:	8334 tonnes (8203 tons)
Dimensions:	162.5m x 16.6m x 7.6m (533ft 2in x 54ft 6in x 25ft)
Machinery:	Twin screws, turbines
Top speed:	32.7 knots
Main armament:	Two 20mm (0.8in) Vulcan guns, two quad Harpoon launchers, two twin launchers for Standard SM-2 ER missiles
Launched:	June 1962

TIMELINE   1958   1962   1976

Mississippi

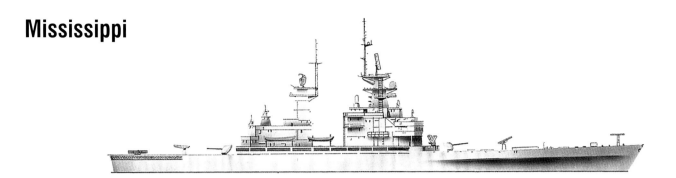

SPECIFICATIONS	
Type:	US cruiser
Displacement:	11,176 tonnes (11,000 tons)
Dimensions:	178.5m x 19.2m x 9m (585ft 4in x 63ft x 29ft 6in)
Machinery:	Twin screws, nuclear-powered turbines
Main armament:	Two 127mm (5in) guns, two twin launchers for Tartar and Harpoon missiles, Asroc launcher
Launched:	July 1976

Mississippi was one of four ships with Mk 26 missile launchers and provision for a helicopter hangar and elevator in the stern. The hangars were replaced by three Tomahawk launchers in the 1980s. The 127mm (5in) Mk 45 guns fired 20 rounds per minute, with a range of over 14.6km (9.12 miles). It was broken up in 1997.

Ticonderoga

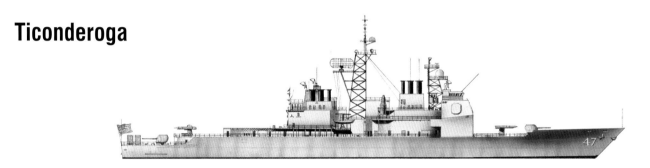

SPECIFICATIONS	
Type:	US guided missile cruiser
Displacement:	9052.5 tonnes (8910 tons)
Dimensions:	171.6m x 19.81m x 9.45m (563ft x 65ft x 31ft)
Machinery:	Twin screws, four gas turbines
Top speed:	30 knots
Main armament:	Eight Harpoon SSM, two Mk 26 launchers for SAM and ASROC torpedoes, two 127mm (5in) guns, six 324mm (12.75in) torpedo tubes
Complement:	343
Launched:	1981

Ticonderoga was lead ship of a class of 27, originally frigates but reclassified due to the scale of their armament. Later vessels in the class were enlarged to carry greater missile stocks. A full array of sensor equipment was carried, including the Aegis air defence system, and one helicopter. It was decommissioned in 2004.

Bunker Hill

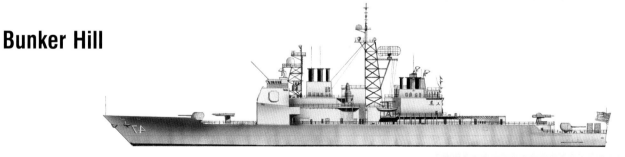

SPECIFICATIONS	
Type:	US guided missile cruiser
Displacement:	9,754 tonnes (9600 tons)
Dimensions:	173m x 16.8m x 10.2m (567ft x 55ft x 34ft)
Machinery:	Twin reversible-pitch screws, four gas turbines
Top speed:	32.5 knots
Main armament:	Two 61-cell Mk 41 VLS systems, 122 RIM-156 SM-2ER Bock IV, RIM-162 ESSM, BGM-109 Tomahawk or RUM-139 VL Asroc; 8 RGM-84 Harpoon missiles
Complement:	400
Launched:	1985

First ship of the large *Ticonderoga* class to be equipped with the Mk 41 Vertical Launching System, *Bunker Hill* was first deployed in the Persian Gulf in 1987, and later participated in the Desert Shield and Desert Storm operations. Weapons systems were upgraded in 2006. In 2010 the ship assisted with disaster relief after the Haiti earthquake.

1981    1985

Kirov

Kirov's superstructure supports radars and early-warning antennae. Most missile-launching systems are forward, below deck, leaving the aft section for the helicopter hangar and machinery. The two nuclear reactors are coupled with oil-fired turbine superheaters to intensify the heat of the steam, increasing power for more speed.

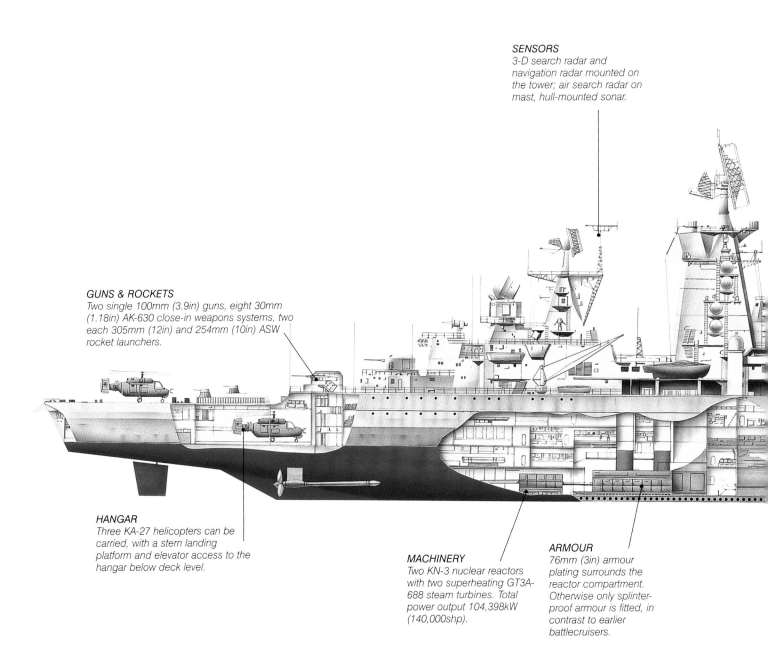

SENSORS
3-D search radar and navigation radar mounted on the tower; air search radar on mast, hull-mounted sonar.

GUNS & ROCKETS
Two single 100mm (3.9in) guns, eight 30mm (1.18in) AK-630 close-in weapons systems, two each 305mm (12in) and 254mm (10in) ASW rocket launchers.

HANGAR
Three KA-27 helicopters can be carried, with a stern landing platform and elevator access to the hangar below deck level.

MACHINERY
Two KN-3 nuclear reactors with two superheating GT3A-688 steam turbines. Total power output 104,398kW (140,000shp).

ARMOUR
76mm (3in) armour plating surrounds the reactor compartment. Otherwise only splinter-proof armour is fitted, in contrast to earlier battlecruisers.

Kirov

SPECIFICATIONS	
Type:	Soviet guided-missile cruiser
Displacement:	28,448 tonnes (28,000 tons)
Dimensions:	248m x 28m x 8.8m (813ft 8in x 91ft 10in x 28ft 10in)
Machinery:	Twin screws, turbines, two pressurized water-type reactors
Top speed:	32 knots
Main armament:	Two 100mm (3.9in) guns, two twin SA-N-4 launchers, twelve SA-N-6 launchers plus 20 anti-ship missiles
Complement:	1600
Launch date:	December 1977

This ship had an impressive armament of missiles and guns as well as electronics. Its largest radar antenna was mounted on its foremast. *Kirov* suffered a reactor accident in 1990 while serving in the Mediterranean Sea. Repairs were never carried out, due to lack of funds and the changing political situation in the Soviet Union.

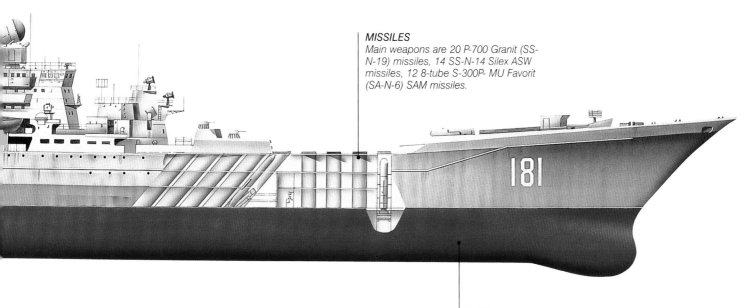

MISSILES
Main weapons are 20 P-700 Granit (SS-N-19) missiles, 14 SS-N-14 Silex ASW missiles, 12 8-tube S-300P- MU Favorit (SA-N-6) SAM missiles.

HULL
The remaining ships of the Kirov (now renamed Admiral Ushakov) class are the world's largest non-carrier warships.

Soviet Cruisers

The Soviet Union built up a substantial cruiser fleet in the 1950s, although building programmes were cut back from the original numbers. Despite efforts to modernize some of them, they were of diminishing strategic value. By later in the century, they were largely obsolescent and were retired or scrapped. Only a few remain active.

Admiral Senyavin

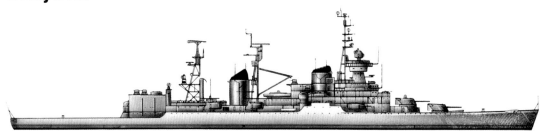

A *Sverdlov*-class heavy cruiser, this ship was extensively modified in 1971–72 to become a command ship in the event of nuclear war, and the rear guns were removed to install a helicopter deck and hangar. An SA-N-4 SAM launcher was fitted at the same time. It served with the Soviet fleet until stricken in 1991.

SPECIFICATIONS	
Type:	Soviet cruiser
Displacement:	16,723 tonnes (16,640 tons)
Dimensions:	210m x 22m x 6.9m (672ft 4in x 22ft 7in x 22ft)
Machinery:	Not known
Top speed:	32.5 knots
Main armament:	Twelve 152mm (6in) guns, ten 533mm (21in) torpedo tubes
Complement:	1250
Launched:	1952

Dmitry Pozharsky

Unlike *Admiral Senyavin*, this ship, sixth of the *Sverdlov* class to be built, retained its original (and increasingly out-of-date) armament. Sixteen ships of this class were built, out of a projected 30. Though obsolete, *Dmitry Pozharsky* remained on active service until 1987. Others were taken out of service 20 years earlier.

SPECIFICATIONS	
Type:	Soviet cruiser
Displacement:	16,906.3 tonnes (16,640 tons)
Dimensions:	210m x 22m x 6.9m (672ft 4in x 22ft 7in x 22ft
Machinery:	Not known
Top speed:	32.5 knots
Main armament:	Twelve 152mm (6in) guns, twelve 100mm (3.9in) guns, ten 533mm (21in) torpedo tubes
Complement:	1250
Launched:	1953

TIMELINE

1952 1953

Dmitri Donskoi

Of the 24 planned ships of this class, 17 were launched, but by the end of 1960 only 14 were operational. Nearly all were fitted for minelaying, mine stowage being on the main deck. The 152mm (6in) guns were mounted in triple turrets, two fore and two aft, each group having its own range-finders.

SPECIFICATIONS

Type:	Soviet cruiser
Displacement:	19,507 tonnes (19,200 tons)
Dimensions:	210m x 21m x 7m (689ft x 70ft x 24ft 6in)
Machinery:	Twin screws, turbines
Main armament:	Twelve 152mm (6in) guns
Launched:	1953

Kerch

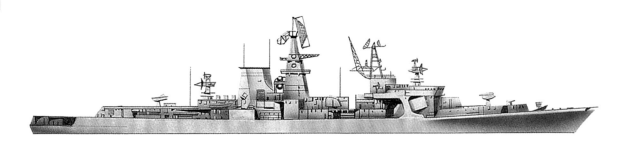

The largest vessels with all gas-turbine propulsion, the *Kerch* class developed 120,000hp, for a range of 5700km (3000 miles) at full speed, or 16,720km (8800 miles) at 15 knots. They had major AA and anti-submarine capabilities, a heavy gun armament, and a helicopter. *Kerch* remains in the Black Sea Fleet.

SPECIFICATIONS

Type:	Soviet cruiser
Displacement:	9855 tonnes (9700 tons)
Dimensions:	173m x 18.6m x 6.7m (567ft 7in x 61ft x 22ft)
Machinery:	Twin screws, gas turbines
Main armament:	Four 76.2mm (3in) guns, two twin SA-N- 3 missile launchers
Complement:	525
Launched:	1973

Slava

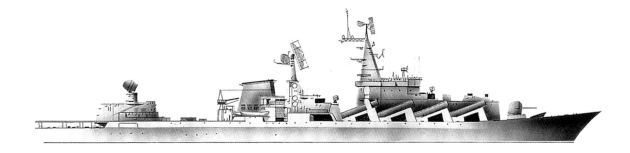

Slava is one of four ships, smaller versions of the *Kirov* class. The SS-N-12 missiles are in twin launchers along each side of the bridge. Complex radars are fitted on the massive foremast at the end of the bridge, with another mounted on top of the mainmast. *Slava* was in the South Ossetia war of 2008.

SPECIFICATIONS

Type:	Soviet cruiser
Displacement:	11,700 tonnes (11,200 tons)
Dimensions:	186m x 20.8m x 7.6m (610ft 5in x 68ft 5in x 25ft)
Machinery:	Twin screws, gas turbines
Main armament:	Two 127mm (5in) guns, eight twin SS-N-12 launchers, eight launchers for SA-N-6 missiles plus two twin launchers for SAM missiles
Launched:	1979

1973  1979

Cruisers of Other Navies

The traditional 'big' surface warship lingered on through the second half of the twentieth century, encouraged by a sense of national prestige and by the desire of admirals to have flagships. Attitudes changed, particularly perhaps after the sinking of the Argentinian *General Belgrano* by the British submarine *Conqueror* in 1982.

Babur

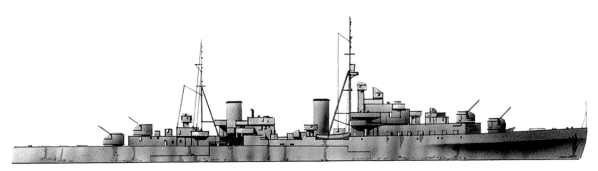

The British cruiser *Diadem* was acquired by Pakistan in 1956 and comprehensively refitted in 1956–57. In 1961 it became a naval training ship, and rarely went to sea, later becoming harbour-bound. In 1982 it was renamed *Jahanqir* and a former British *Devonshire* class destroyer became *Babur*. It was broken up in 1985.

SPECIFICATIONS	
Type:	Pakistani cruiser
Displacement:	7638.3 tonnes (7518 tons)
Dimensions:	156m x 15m x 5.4m (512ft x 50ft 6in x 18ft)
Machinery:	Quadruple screws, geared tubines
Top speed:	32 knots
Main armament:	Eight 133mm (5.25in) guns
Armour:	76mm (3in) sides, 51–25mm (2–1in) deck
Complement:	530
Launched:	1942 (refitted 1956)

Coronel Bolognesi

One of two British *Fiji*-class cruisers sold to Peru, this was formerly HMS *Ceylon*. Modified before delivery, it received a new mast, increased anti-aircraft armament and improved bridge accommodation. Radar equipment was also updated, for long-range search, height-finding and fire-control. It remained active until 1982.

SPECIFICATIONS	
Type:	Peruvian cruiser
Displacement:	11,633 tonnes (11,450 tons)
Dimensions:	169.4m x 18.9m x 6.4m (555ft 6in x 62ft x 20ft 9in)
Machinery:	Quadruple screws, geared turbines
Top speed:	31.5 knots
Main armament:	Nine 152mm (6in) guns, eight 102mm (4in) guns
Complement:	920
Launched:	1942 (refitted 1960)

TIMELINE

1956   1960   1957

Prat

SPECIFICATIONS	
Type:	Chilean cruiser
Displacement:	12,405 tonnes (12,210 tons)
Dimensions:	185.4m x 18.8m x 6.95m (608ft 4in x 61ft 9in x 22ft 9in)
Machinery:	Quadruple screws, geared turbines
Top speed:	30 knots
Main armament:	Fifteen 152mm (6in) guns, eight 127mm (5in) guns
Armour:	127mm (5in) belt, 51mm (2in) deck
Complement:	868
Launched:	1936 (modernized 1957)

Formerly the US *Brooklyn* class cruiser *Nashville,* this World War II veteran was transferred to the Chilean Navy in 1951, as was *Brooklyn* itself, renamed *O'Higgins.* Both ships were refitted and modernized in the United States in 1957–58. *Prat* was decommissioned in 1984, while *O'Higgins* was stricken only in 1992.

Tiger

SPECIFICATIONS	
Type:	British cruiser
Displacement:	12,273 tonnes (12,080 tons)
Dimensions:	170m x 20m x 6.4m (555ft 6in x 64ft x 21ft 3in)
Machinery:	Quadruple screws, turbines
Top speed:	31.5 knots
Main armament:	Four 152mm (6in) guns, six 76mm (3in) guns
Launched:	October 1945/refitted 1959

Tiger was originally laid down in 1941, but work stopped in 1946 and it was finally completed to new plans in 1959, as one of the last cruisers to enter British service. However, it proved unsuitable for the conditions and requirements of the period, and after little use, was withdrawn in the 1960s and scrapped in 1986.

Caio Duilio

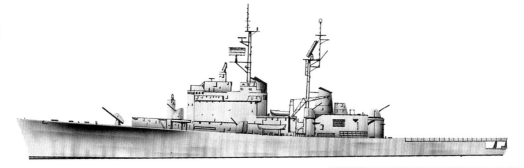

SPECIFICATIONS	
Type:	Italian cruiser
Displacement:	6604 tonnes (6506 tons)
Dimensions:	144m x 17m x 4.7m (472ft 4in x 55ft 7in x 15ft 4in)
Machinery:	Twin screws, geared turbines
Top speed:	31 knots
Main armament:	Eight 76mm (3in) guns, plus Terrier surface-to-air missiles
Complement:	485
Launched:	December 1962

Caio Duilio and its sister-ship *Andrea Doria* were helicopter cruisers for anti-submarine and air defence, designed to a new plan, having a wide beam in relation to their length. They carried three AB 212SW armed helicopters. In 1980, *Caio Duilio* became a training ship, and was decommissioned in 1991.

1959   1962

Destroyers (guided missile): Part 1

The introduction of the guided missile and the new generation of long-range guided or homing torpedoes, as well as radar-directed, rapid-fire guns, made the medium-sized warship a more formidable fighting craft than ever before. Destroyer design, construction and ship-management entered a phase of rapid, significant change.

Farragut

This class became the US Navy's first missile ships, and *Farragut* was among the first to carry the ASROC system, which fired rocket-boosted anti-submarine torpedoes. Later it was fitted with the Naval Tactical Data System for air defence command and control. Decommissioned in 1989, it was stricken in 1992.

SPECIFICATIONS

Type:	US destroyer
Displacement:	5738.4 tonnes (5648 tons)
Dimensions:	156.3m x 15.9m x 5.3m (512ft 6in x 52ft 4in x 17ft 9in)
Machinery:	Twin screws, geared turbines
Top speed:	32 knots
Main armament:	One Terrier (later Standard) SAM missile system, one ASROC rocket-boosted ASW torpedo launcher, six 324mm (12.75in) torpedo tubes
Complement:	360
Launched:	1958

Gremyaschiy

The 'Krupny' class of nine destroyers were the first missile-armed ships in the Soviet fleet. The original SS-N-1 Scrubber missile system was soon removed and the class rearmed for anti-submarine warfare. With 16 57mm (2.24in) guns, the ship was lightly armed for self-defence, reducing its usefulness. It was stricken in 1995.

SPECIFICATIONS

Type:	Soviet destroyer
Displacement:	4259 tonnes (4192 tons)
Dimensions:	138.9m x 14.84m x 4.2m (455ft 9in x 48ft 8in x 13ft 9in)
Machinery:	Twin screws, geared turbines
Top speed:	34.5 knots
Main armament:	Two SSN-N-1 SSM, two anti-submarine rocket launchers, six 533mm (21in) torpedo tubes
Complement:	310
Launched:	1959

TIMELINE

1958   1959   1960

Devonshire

Designed at the end of the 1950s, *Devonshire* and seven others were built to operate in the fall-out area of a nuclear explosion, with deck installations under cover. Later, Seacat missiles replaced the Seaslug and some of the class also carried the Exocet anti-ship missile. *Devonshire* was sunk as a target in 1984.

SPECIFICATIONS	
Type:	British destroyer
Displacement:	6299 tonnes (6200 tons)
Dimensions:	158m x 16m x 6m (520ft 6in x 54ft x 20ft)
Machinery:	Twin screws, turbines plus four gas turbines
Top speed:	32.5 knots
Main armament:	Four 114mm (4.5in) guns, twin launcher for Seaslug missile
Complement:	471
Launched:	June 1960

Boykiy

Boykiy was originally completed as a missile ship armed with SS-N-1 launchers. When these became obsolete in the mid-1960s, it was converted into an anti-submarine vessel. *Boykiy* served in the North Atlantic and the North Pacific. It was towed to Spain for breaking in 1988, grounding off Norway on the way.

SPECIFICATIONS	
Type:	Soviet destroyer
Displacement:	4826 tonnes (4750 tons)
Dimensions:	140m x 15m x 5m (458ft 9in x 49ft 5in x 16ft 6in)
Machinery:	Twin screws, geared turbines
Top speed:	35 knots
Main armament:	Eight 57mm (2.24in) guns, plus missiles
Complement:	310
Launched:	1960

Impavido

Impavido was one of Italy's first two missile-armed destroyers. Derived from the *Impetuoso*-class destroyers, it had a US Mk 13 launcher for Tartar surface-to-air missiles. The after funnel was heightened to clear the fire control tracker on top of the aft structure. Modernized in 1976–77, *Impavido* was decommissioned in 1992.

SPECIFICATIONS	
Type:	Italian destroyer
Displacement:	4054 tonnes (3990 tons)
Dimensions:	131.3m x 13.7m x 4.4m (430ft 9in x 45ft x 14ft 5in)
Machinery:	Twin screws, turbines
Top speed:	34 knots
Main armament:	Two 127mm (5in) guns, one Tartar missile launcher, six 533mm (21in) torpedo tubes
Complement:	340
Launched:	May 1962

1962

Destroyers (guided missile): Part 2

As usual with warship types, destroyers tended to get larger, offering a more substantial launch platform for various missiles. The proposed US *Zumwalt* class, to replace the *Arleigh Burkes* in the years after 2010, has an anticipated displacement of 12,000 tonnes (11,808 tons), equivalent to a heavy cruiser of 60 years ago.

Ognevoy

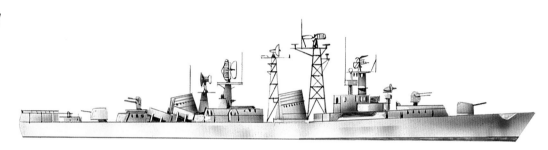

Project 61 from the 1960s produced 20 guided missile destroyers, known as the 'Kashin' class. The third completed, *Ognevoy*, was given cruise missiles and sonar and new air defence in the mid-1970s. Sensor equipment included air-search, navigation and fire-control radar. *Ognevoy* was broken up in 1990.

SPECIFICATIONS	
Type:	Soviet destroyer
Displacement:	4460.3 tonnes (4390 tons)
Dimensions:	144m x 15.8m x 4.6m (472ft 5in x 51ft 10in x 15ft 1in)
Machinery:	Twin screws, four gas turbines
Top speed:	18 knots
Main armament:	Two SA-N-1 SSM launchers, two RBU-6000 and two RBU-1000 anti-submarine rocket launchers, four 76mm (3in) guns, five 533mm (21in) torpedo tubes
Complement:	266
Launched:	1963

Duquesne

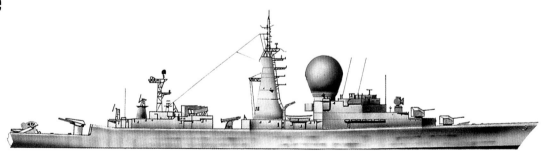

With its sister ship *Suffren*, *Duquesne* was the first French destroyer designed to carry surface-to-air missiles, serving as escort ships for the new generation of French aircraft carriers. *Duquesne* also received four Exocet missile launchers. Electronics were modernized in 1990–91, and it was decommissioned in 2007.

SPECIFICATIONS	
Type:	French destroyer
Displacement:	6187 tonnes (6090 tons)
Dimensions:	157.6m x 15.5m x 7m (517ft x 50ft 10in x 23ft 9in)
Machinery:	Twin screws, turbines
Top speed:	34 knots
Main armament:	Two 100mm (3.9in) guns, one Malafon anti-submarine missile launcher/four torpedo launchers
Launched:	February 1966

TIMELINE　　1963　　　　　1966　　　　　1973

Duguay-Trouin

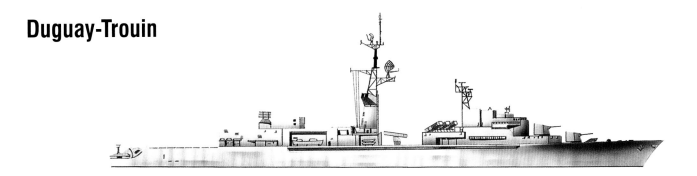

The three *Tourville*-class guided-missile destroyers were the first French warships of destroyer size purpose-built to operate two anti-submarine helicopters. In 1979, *Duguay-Trouin's* Crotale missile launcher replaced a third gun turret, and new generation air-surveillance radar was fitted. It was decommissioned in 1999.

SPECIFICATIONS	
Type:	French destroyer
Displacement:	5892 tonnes (5800 tons)
Dimensions:	152.5m x 15.3m x 6.5m (500ft 4in x 50ft 2in x 21ft 4in)
Machinery:	Twin screws, turbines
Top speed:	32 knots
Main armament:	Two 100mm (3.9in) guns, one eight-cell Crotale launcher
Complement:	282
Launched:	June 1973

Spruance

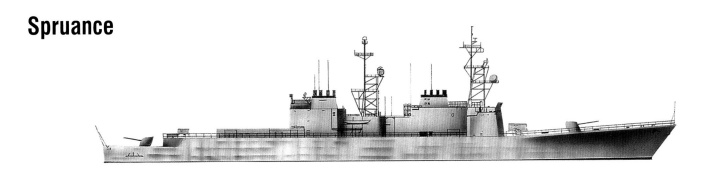

Bigger than many former cruisers, the *Spruance* class was designed to offer a stable weapon-launcher, able to operate in difficult sea conditions. Its successful hull design was used, with modifications, on two other classes of US warship. After serving in the Atlantic fleet, *Spruance* was sunk as a target in 2006.

SPECIFICATIONS	
Type:	US destroyer
Displacement:	8168 tonnes (8040 tons)
Dimensions:	171.7m x 16.8m x 5.8m (563ft 4in x 55ft 2in x 19ft)
Machinery:	Twin screws, gas turbines
Top speed:	32.5 knots
Main armament:	Two 127mm (5in) guns, Tomahawk and Harpoon missiles
Complement:	296
Launched:	1973

Tachikaze

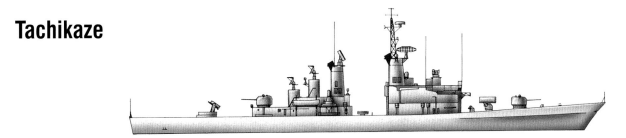

Like all Japanese post-war navy ships, *Tachikaze* was equipped with American weapons. Two 20mm (0.79in) Phalanx CIWS were added in 1983, when the Harpoon SSM were installed. It could also accommodate a SH-601 helicopter. From 1998 to decommissioning in January 2007, it was the Fleet Escort Force flagship.

SPECIFICATIONS	
Type:	Japanese destroyer
Displacement:	4877 tonnes (4800 tons)
Dimensions:	143m x 14.3m x 4.6m (469ft 2in x 46ft 10in x 15ft 1in)
Machinery:	Twin screws, geared turbines
Top speed:	32 knots
Main armament:	Eight Harpoon SSM, one Mk 13 Standard SAM launcher, one ASROC launcher, two 127mm (5in) guns
Complement:	277
Launched:	1974

 1974

Destroyers (guided missile): Part 3

Air and surface combat are usually seen as the prime role of the destroyer, though some navies have also fitted out destroyers for anti-submarine roles. CIWS (close-in weapons system) response systems are generally fitted, based on 20mm or 30mm (0.79in or 1.18in) guns, with a variety of missile and anti-missile systems, depending on the supplier.

Santisima Trinidad

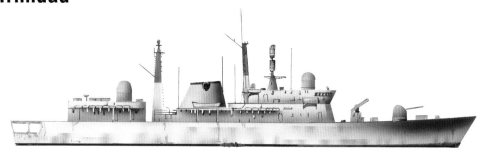

Based on the Royal Navy's Type 42 destroyer, and carrying a Lynx helicopter, *Santisima Trinidad* was commissioned in 1981. In April 1982, it was the lead ship in Argentina's invasion of the Falkland Islands. In the late 1980s and 1990s, much of its equipment was cannibalized to keep its sister ship *Hercules* active.

SPECIFICATIONS

Type:	Argentinian destroyer
Displacement:	4419.6 tonnes (4350 tons)
Dimensions:	125m x 14m x 5.8m (410ft x 46ft x 19ft)
Machinery:	Twin screws, four gas turbines
Top speed:	30 knots
Main armament:	One GWS30 Sea Dart SAM missile launcher, one 114mm (4.5in) gun, six 324mm (12.5in) torpedo tubes
Complement:	312
Launched:	1974

Dupleix

Dupleix was one of eight ships built at Brest as anti-submarine vessels. A major innovation was the use of gas turbine engines, developing 52,000hp for a speed of 30 knots. All in the class had a double hangar aft for helicopters. The ships are well adapted for surface work, as well as the standard ASW role.

SPECIFICATIONS

Type:	French destroyer
Displacement:	4236 tonnes (4170 tons)
Dimensions:	139m x 14m x 5.7m (456ft x 46ft x 18ft 8in)
Machinery:	Twin screws, gas turbines, plus diesels
Top speed:	30knots
Main armament:	One 100mm (3.9in) gun, four MM38 Exocet SSM launchers, octuple Crote Navale SAM launcher, two fixed torpedo launchers
Complement:	216
Launched:	December 1978

TIMELINE

1974　　1978　　1982

Euro

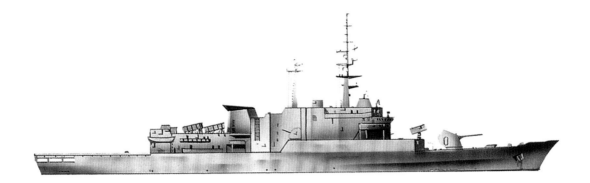

Euro entered service in 1983. The aft flight deck is 27m (88ft 6in) long and 12m (39ft 4in) wide, and a Variable Depth Sonar is streamed out on a 900m (984yd) long cable from a stern well. As the Italian Navy does not use the term 'destroyer', it is classed as a guided-missile frigate.

SPECIFICATIONS	
Type:	Italian destroyer
Displacement:	3088 tonnes (3040 tons)
Dimensions:	122.7m x 12.9m x 8.4m (402ft 6in x 42ft 4in x 27ft 6in)
Machinery:	Twin screws, diesels and gas turbines
Top speed:	29 knots (diesel), 32 knots (turbines)
Main armament:	One 127mm (5in) gun plus missiles
Launched:	December 1982

Mutenia

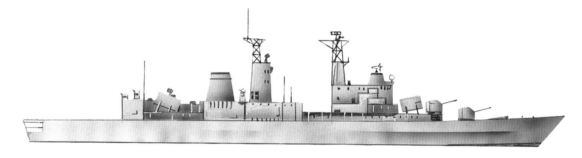

Built in Romania to Russian designs and using mostly Russian equipment, *Mutenia* was well equipped, having two Alouette III helicopters. It spent lengthy periods out of service, suffering from the problems of a one-off type as well as a lack of operating funds. In 1990–92, it was refitted with anti-submarine weapons.

SPECIFICATIONS	
Type:	Romanian destroyer
Displacement:	5882.6 tonnes (5790 tons)
Dimensions:	144.6m x 14.8m x 7m (474ft 4in x 48ft 6in x 23ft)
Machinery:	Twin screws, diesels
Top speed:	31 knots
Main armament:	Eight SSN-N-2C SSM, four 76mm (3in) guns, six 533mm (21in) torpedo tubes, two RBU-120 anti-submarine rocket launchers
Complement:	270
Launched:	1982

Hamayuki

Japan's Maritime Self Defence Force produced some versatile, well-armed ships in the 1980s. *Hamayuki* was fifth of 12 *Hatsuyuki* class ships and carried a Mitsubishi-built HSS 2B Sea King helicopter. Sensor equipment includes hull-mounted sonar, and air-search, sea-search and fire-control radar.

SPECIFICATIONS	
Type:	Japanese destroyer
Displacement:	3759.2 tonnes (3700 tons)
Dimensions:	131.7m x 13.7m x 4.3m (432ft 4in x 44ft 11in x 14ft 3in)
Machinery:	Twin screws, four gas turbines
Top speed:	30 knots
Main armament:	Eight Harpoon SSM, one Sea Sparrow SAM, one 76mm (3in) gun, two Mk15 Phalanx 20mm (0.79in) CIWS, one ASROC, six 324mm (12.7in) torpedo tubes
Complement:	190
Launched:	1983

1983

Destroyers (guided missile): Part 4

Gas turbine propulsion, enabling rapid acceleration to maximum speed, is favoured for modern destroyers. The Soviet 'Kashin' class used gas turbines in the 1960s, as did the Canadian *Iroquois* class of the 1970s and the ships of the US *Spruance* class, launched between 1972 and 1980, and the *Arleigh Burke* class, still in service.

Edinburgh

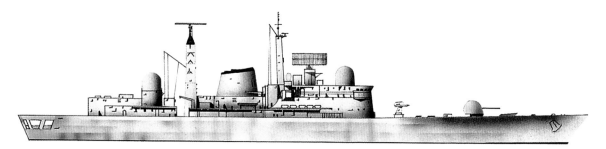

Edinburgh was designed as an air-defence/ASW ship to work with a naval or amphibious task force. Its two helicopters carry air-to-surface weapons for use against lightly defended surface ships. Refitted in 1990, *Edinburgh* saw service off Iraq in the second Gulf War (2003) and had a further full refit in 2004–5.

SPECIFICATIONS	
Type:	British destroyer
Displacement:	4851 tonnes (4775 tons)
Dimensions:	141m x 14.9m x 5.8m (463ft x 48ft x 19ft)
Machinery:	Twin screws, gas turbines
Top speed:	30 knots
Main armament:	One 114mm (4.5in) gun, helicopter-launched Mk44 torpedoes, two triple mounts for Mk46 anti-submarine torpedoes, one Sea Dart launcher
Launched:	March 1983

Arleigh Burke

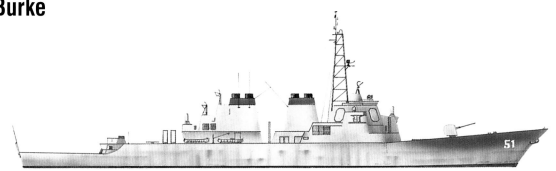

This class was designed to replace the *Adams* and *Coontz* class destroyers, which entered service in the early 1960s. *Arleigh Burke* was commissioned in 1991, to provide effective anti-aircraft cover, for which the SPY 1 D version of the Aegis system was fitted. It also has anti-surface and anti-submarine weapons.

SPECIFICATIONS	
Type:	US guided missile destroyer
Displacement:	8534 tonnes (8400 tons)
Dimensions:	142.1m x 18.3m x 9.1m (266ft 3in x 60ft x 30ft)
Machinery:	Twin screws, gas turbines
Top speed:	30+ knots
Main armament:	Harpoon and Tomahawk missiles, 127mm (5in) gun
Launched:	1989

TIMELINE   1983   1989   1991

Kongo

SPECIFICATIONS	
Type:	Japanese destroyer
Displacement:	9636.8 tonnes (9485 tons)
Dimensions:	160.9m x 20.9m x 6.2m (520ft 2in x 68ft 7in x 20ft 4in)
Machinery:	Twin screws, gas turbines
Top speed:	30 knots
Main armament:	Eight Harpoon SSM, two Mk41 VLS with Standard missiles and ASROC torpedoes, one 127mm (5in) gun, two 20mm (0.79in) Phalanx CIWS, six 324mm (12.75in) torpedo tubes
Complement:	300
Launched:	1991

Kongo is based on the US *Arleigh Burke* class, its dimensions easily justifying cruiser designation. The flight deck has no hangar but can take a SH-60J Seahawk helicopter. Fitted with the Aegis air defence radar and missile system, the class has recently been modified to intercept North Korean ballistic missiles.

Brandenburg

SPECIFICATIONS	
Type:	German destroyer
Displacement:	4343.4 tonnes (4275 tons)
Dimensions:	138.9m x 16.7m x 6.3m (455ft 8in x 57ft 1in x 20ft 8in)
Machinery:	Twin screws, gas turbines plus diesels
Top speed:	29 knots
Main armament:	Four MM38 Exocet SSM, one VLS for Sea Sparrow SAM, two 21-cell RAM launchers, one 76mm (3in) gun, six 324mm (12.75in) torpedo tubes
Complement:	219
Launched:	1992

Brandenburg was lead ship of a class of four, identified as Type 123, and intended for air defence. They are equipped with air-search and air/surface search radar, two fire-control trackers and hull-mounted sonar. Two Lynx Mk88 helicopters are carried. Germany has built similar ships for Portugal and Turkey.

Murasame

SPECIFICATIONS	
Type:	Japanese destroyer
Displacement:	5181.6 tonnes (5100 tons)
Dimensions:	151m x 16.9m x 5.2m (495ft 5in x 55ft 7in x 17ft 1in)
Machinery:	Twin screws, gas turbines
Top speed:	33 knots
Main armament:	Eight Harpoon SSM, two Mk41 VLS with Standard missiles and ASROC torpedoes, one 127mm (5in) gun, two 20mm (0.79in) Phalanx CIWS, six 324mm (12.75in) torpedo tubes
Complement:	170
Launched:	1994

Commissioned in 1996, designed primarily for air defence but with anti-submarine potential, *Murasame* is a typically all-round member of the modern Japanese fleet. An SH-60J helicopter is carried, with hangar. The *Murasame* class has full radar search equipment, and both hull-mounted and towed sonar.

1992   1994

Destroyers (anti-submarine): Part 1

From its inception as a warship-type, torpedoes were the main armament of a destroyer, enabling it to be a threat to much larger ships. While many destroyers still carry torpedo tubes, these no longer define the destroyer as such. Some ships classed as destroyers have no torpedoes, relying on missiles and and ASW mortars.

St Laurent

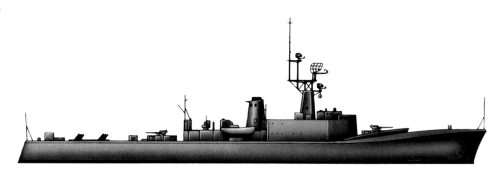

Seven ships formed the *St Laurent* class, the lead vessel being completed in 1955. In the early 1960s, the armament was updated and variable-depth sonar was mounted at the stern. A helicopter deck and hangar were built in. *St Laurent* was decommissioned in 1979 and sank under tow to the breakers in 1980.

SPECIFICATIONS	
Type:	Canadian destroyer
Displacement:	2641.6 tonnes (2600 tons)
Dimensions:	111.6m x 12.8m x 4m (366ft x 42ft x 13ft 2in)
Machinery:	Twin screws, turbines
Top speed:	28 knots
Main armament:	Four 76mm (3in) guns, two Limbo Mk10 anti-submarine mortars
Complement:	290
Launched:	1951

Neustrashimyy

Planned as lead ship of a large class, completed in 1955, *Neustrashimyy* ended up as a one-off. But it displayed many features to reflect the nuclear era, including air-conditioning and sealable crew accommodation. Its pressure-fired boiler design was used in many subsequent ships. It was broken up in 1975.

SPECIFICATIONS	
Type:	Soviet destroyer
Displacement:	3434 tonnes (3830 tons)
Dimensions:	133.8m x 13.6m x 4.4m (439ft 1in x 44ft 6in x 14ft 6in)
Machinery:	Twin screws, geared turbines
Top speed:	36 knots
Main armament:	Four 130mm (5in) guns, 10 533mm (21in) torpedo tubes
Complement:	305
Launched:	1951

TIMELINE

 1951  1952

Grom

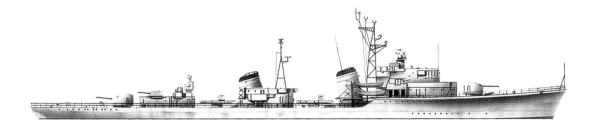

Grom was the former Soviet *Smetlivy*, one of two destroyers transferred from the USSR Baltic Fleet to Poland in 1957. It was a member of a class comprising the first Russian destroyers built after World War II, and incorporating features from German destroyers. In service until 1973, *Grom* was scrapped in 1977.

SPECIFICATIONS	
Type:	Polish destroyer
Displacement:	3150 tonnes (3100 tons)
Dimensions:	120.5m x 11.8m x 4.6m (395ft 4in x 38ft 9in x 15ft)
Machinery:	Twin screws, turbines
Main armament:	Four 130mm (5.1in) guns, two 76mm (3in) anti-aircraft guns
Launched:	1952

Groningen

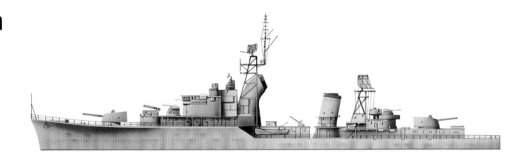

Groningen's class of eight had some side armour as well as deck protection. They were also some of the first destroyers to have no torpedo capability. Two short-range ASW rocket launchers were fitted. It was one of seven of the class sold to Peru in the 1980s. As *Galvez*, it was deleted in 1991.

SPECIFICATIONS	
Type:	Dutch destroyer
Displacement:	3119 tonnes (3070 tons)
Dimensions:	116m x 11.7m x 3.9m (380ft 3in x 38ft 6in x 13ft)
Machinery:	Twin screws, turbines
Top speed:	36 knots
Main armament:	Four 120mm (4.7in) guns
Launched:	January 1954

Almirante Riveros

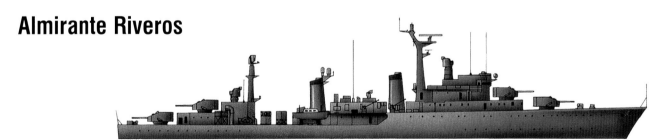

This heavily-armed destroyer was one of a pair built in England. It returned to the UK for modernization work in 1975, when missile systems were fitted, replacing its secondary 40mm (1.57in) armament, and Squid anti-submarine mortars. It served for another 20 years. Decommissioned in 1998, it was sunk as a target in that year.

SPECIFICATIONS	
Type:	Chilean destroyer
Displacement:	3650 tonnes (3300 tons)
Dimensions:	122.5m x 13.1m x 4m (402ft x 43ft x 13ft 4in)
Machinery:	Twin screws, turbines
Top speed:	34.5 knots
Main armament:	Four 102mm (4in) guns, four MM38 Exocet SSM, one Seacat SAM system, six 324mm (12.75in) torpedo tubes
Complement:	266
Launched:	December 1958

1954   1958

Destroyers (anti-submarine): Part 2

Destroyer-borne ASW helicopters are fitted with sonobuoys, dipping sonar and magnetic anomaly detectors to identify potential submerged targets, and armed with torpedoes or depth charges. Helicopters have become of such combat value that some ships are identified as 'helicopter destroyers' or 'helicopter cruisers'.

Coronel Bolognesi

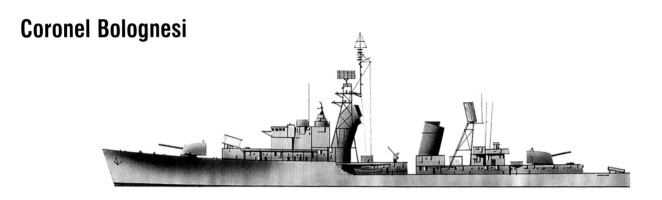

Coronel Bolognesi was formerly the Dutch Overijssel, of the Friesland class. Between 1980 and 1982, they were transferred to the Peruvian Navy, and fitted with Exocet missiles and other new weapons systems and sensors. Coronel Bolognesi arrived in Peru in July 1982 and was decommissioned in 1990.

SPECIFICATIONS	
Type:	Peruvian destroyer
Displacement:	3150 tonnes (3100 tons)
Dimensions:	116m x 12m x 5m (380ft 7in x 38ft 5in x 17ft)
Machinery:	Twin screws, turbines
Top speed:	36 knots
Main armament:	Four 120mm (4.7in) guns
Launched:	August 1955

Aragua

British-built, Aragua was one of the three Nueva Esparta class destroyers. Its two sister-ships were later given more up-to-date armament, including Seacat surface-to-air missiles, and their radar systems were also modernized. Aragua remained very much as delivered. It was withdrawn from service in 1975.

SPECIFICATIONS	
Type:	Venezuelan destroyer
Displacement:	3353 tonnes (3300 tons)
Dimensions:	122.5m x 13.1m x 3.9m (402ft x 43ft x 12ft 9in)
Machinery:	Twin screws, geared turbines
Top speed:	34.5 knots
Main armament:	Six 114mm (4.5in) guns, three 533mm (21in) torpedo tubes, two Squid anti-submarine mortars, two depth charge racks
Complement:	254
Launched:	1955

TIMELINE

1955   1966

Asagumo

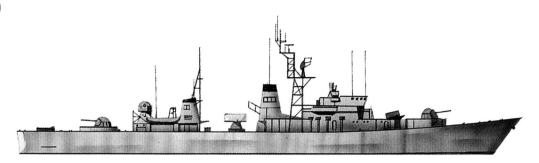

Asagumo and five sister-ships were typical of the mid period of Japanese destroyers after World War II. With 711 tonnes (700 tons) of oil fuel, *Asagumo* had a range of 11,400km (6000 miles) at 20 knots. The gunnery, radar and sensors were all supplied by the United States. It was decommissioned in 1998.

SPECIFICATIONS	
Type:	Japanese destroyer
Displacement:	2083 tonnes (2050 tons)
Dimensions:	114m x 11.8m x 4m (374ft x 38ft 9in x 13ft)
Machinery:	Twin screws, diesel engines
Main armament:	Four 76mm (3in) guns, six torpedo tubes
Launched:	1966

Audace

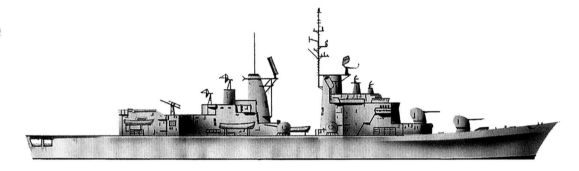

Audace was a multi-function fleet escort, its prime task being anti-submarine action. Two helicopters with weapons kit and sensors were carried. Some of its weapons had a poor arc of fire due to the height of the superstructure. Serving off Lebanon in 1982 and in the Gulf in 1990–91, it was decommissioned in 2006.

SPECIFICATIONS	
Type:	Italian destroyer
Displacement:	4470 tonnes (4400 tons)
Dimensions:	135.9 x 14.6m x 4.5m (446ft x 48ft x 15ft)
Machinery:	Twin screws, geared turbines
Top speed:	33 knots
Main armament:	Two 127mm (5in) guns, one SAM launcher
Launched:	1971

Haruna

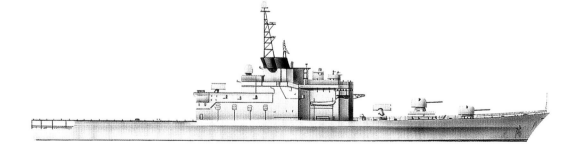

Haruna is a command ship for anti-submarine escort groups, its entire aft part devoted to facilities for three Sea King helicopters. The hangar occupies the full beam of the vessel. Automatic guns fire up to 40 rounds per minute. In 1986–87, *Haruna* underwent a major refit to improve its anti-aircraft defences.

SPECIFICATIONS	
Type:	Japanese destroyer
Displacement:	5029 tonnes (4950 tons)
Dimensions:	153m x 17.5m x 5.2m (502ft x 57ft 5in x 17ft)
Machinery:	Twin screws, turbines
Main armament:	Two 127mm (5in) guns, Sea Sparrow missile launcher, six 324mm (12.75in) torpedo tubes
Launched:	February 1972

1971    1972

Destroyers & Frigates (anti-submarine)

Around 200 crew are needed for a destroyer – not a large number in relation to the size of the ship and its firepower, and made possible by the use of technology. Post-2010 destroyers will include a 3-D phased array radar system in their equipment.

Gurkha

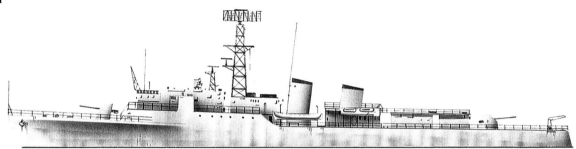

Gurkha was the third of seven all-purpose frigates of the 'Tribal' class, among the first British warships to be air-conditioned in all crew areas and most working spaces. Decommissioned in 1979, reactivated at the time of the Falklands conflict in 1982, *Gurkha* was sold to Indonesia in 1985, and laid up in 1999.

SPECIFICATIONS	
Type:	British frigate
Displacement:	2743 tonnes (2700 tons)
Dimensions:	109m x 12.9m x 5.3m (360ft x 42ft 4in x 17ft 6in)
Machinery:	Single screw, turbine and gas turbine
Top speed:	28 knots
Main armament:	Two 114mm (4.5in) guns, one Limbo three-barrelled anti-submarine mortar
Complement:	253
Launched:	July 1960

Georges Leygues

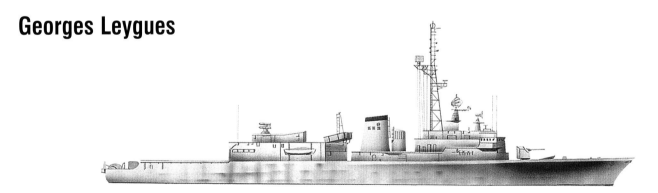

Georges Leygues and its seven sister destroyers are France's prime anti-submarine force. Gas turbine engines develop 52,000hp, while diesels develop 10,400hp; the cruising range at 18 knots on diesels is 18,050km (9500 miles). *Georges Leygues* carries two Lynx helicopters and has full hangar facilities.

SPECIFICATIONS	
Type:	French destroyer
Displacement:	4236 tonnes (4170 tons)
Dimensions:	139m x 14m x 5.7m (456ft x 46ft x 18ft 8in)
Machinery:	Twin screws, gas turbines and diesel engines
Top speed:	30 knots
Main armament:	One 100mm (3.9in) gun, Exocet missiles
Launched:	December 1976

TIMELINE

1960   1976   1978

Glasgow

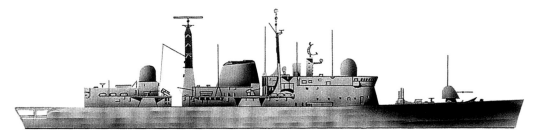

Glasgow was active in the Falklands War, and served at East Timor and in the South Atlantic patrol. It had air-search radar, and a fire control system, and carried one helicopter. Two triple 324mm (12.75in) anti-submarine torpedo tube sets were fitted. Decommissioned in 2005, it was broken up in Turkey in 2009.

SPECIFICATIONS	
Type:	British destroyer
Displacement:	4165 tonnes (4100 tons)
Dimensions:	125m x 14.3m x 5.8m (410ft x 47ft x 19ft)
Machinery:	Twin screws, gas turbines
Top speed:	30 knots
Main armament:	One 114mm (4.5in) gun, one twin Sea Dart mount
Launched:	April 1976

Cushing

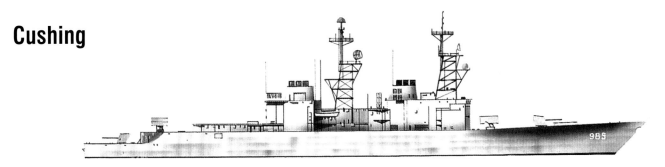

Cushing was the last survivor of the Spruance class, sunk as a target in 2005. Anti-submarine ships, they carried two helicopters, plus the Phalanx CIWS air defence system, and Harpoon and Sparrow missiles. The first US Navy surface vessels fitted with gas turbines, they could run on a single engine at 19 knots.

SPECIFICATIONS	
Type:	US destroyer
Displacement:	7924 tonnes (7800 tons)
Dimensions:	161m x 17m x 9m (529ft 2in x 55ft 1in x 28ft 10in)
Machinery:	Twin screws, gas turbines
Top speed:	30 knots
Main armament:	Two 127mm (5in) guns, six 322mm (12.75in) torpedo tubes
Launched:	June 1978

Hatsuyuki

A radical departure from previous Japanese anti-submarine destroyer designs, Hatsuyuki resembles the French Georges Leygues class in its layout, although its weapons systems are of US origin. The propulsion machinery is British: two groups of gas turbines, one set developing 56,780hp, and the other 10,680hp.

SPECIFICATIONS		
Type:	Japanese destroyer	
Displacement:	3760 tonnes (3700 tons)	
Dimensions:	131.7m x 13.7m x 4.3m (432ft x 45ft x 14ft)	
Machinery:	Twin screws, gas turbines	
Top speed:	30 knots	
Main armament:	One 76mm (3in) gun, one eight-cell Sea Sparrow launcher, two 20mm (0.79in) Phalanx CIWS	
Complement:	190	
Launched:	November 1980	

1978

1980

Tromp

Tromp and sister-ship *De Ruyter* acted as flagships to two long-range NATO task groups, for operating in the Eastern Atlantic. An octuple Sea Sparrow launcher with 60 reloads provided short-range anti-aircraft and anti-missile defence, and a later refit added a Goalkeeper point defence gun system. A single Lynx helicopter is carried.

Tromp

Replacing two cruisers in service with the Royal Netherlands navy, HNLMS *Tromp* and *De Ruyter* were among the largest and most capable of frigates afloat. Weapons fitted included Harpoon, Standard and Sea Sparrow missiles.

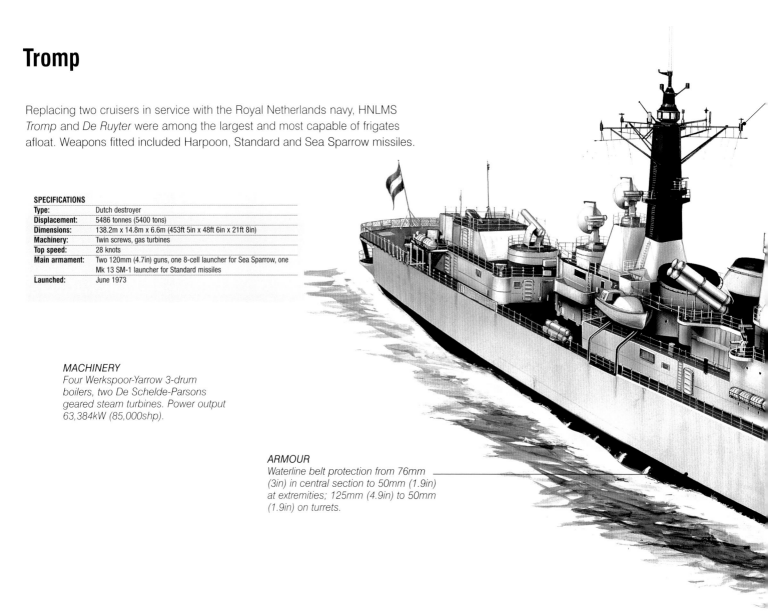

SPECIFICATIONS

Type:	Dutch destroyer
Displacement:	5486 tonnes (5400 tons)
Dimensions:	138.2m x 14.8m x 6.6m (453ft 5in x 48ft 6in x 21ft 8in)
Machinery:	Twin screws, gas turbines
Top speed:	28 knots
Main armament:	Two 120mm (4.7in) guns, one 8-cell launcher for Sea Sparrow, one Mk 13 SM-1 launcher for Standard missiles
Launched:	June 1973

MACHINERY
Four Werkspoor-Yarrow 3-drum boilers, two De Schelde-Parsons geared steam turbines. Power output 63,384kW (85,000shp).

ARMOUR
Waterline belt protection from 76mm (3in) in central section to 50mm (1.9in) at extremities; 125mm (4.9in) to 50mm (1.9in) on turrets.

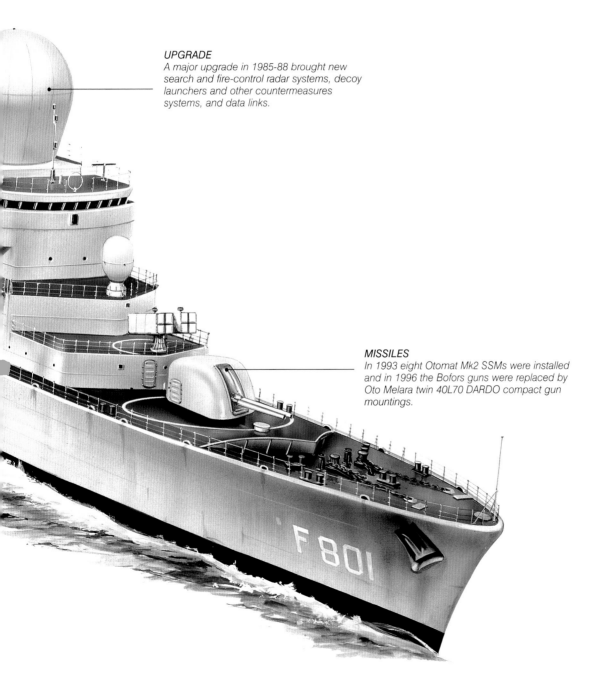

MASTS
In original form the ship was unusual in having no separate masts, with extensions to the control tower and funnel serving instead.

UPGRADE
A major upgrade in 1985-88 brought new search and fire-control radar systems, decoy launchers and other countermeasures systems, and data links.

MISSILES
In 1993 eight Otomat Mk2 SSMs were installed and in 1996 the Bofors guns were replaced by Oto Melara twin 40L70 DARDO compact gun mountings.

F 801

Frigates of the 1950s

It is not always easy to attach type-names to modern warships. 'Frigate' is a case in point. Earlier in the twentieth century, this described an escort ship, especially for merchant convoys. It was generally smaller and slower than a destroyer, and not armed with torpedoes. In the 1950s, this description was already ceasing to fit.

Grafton

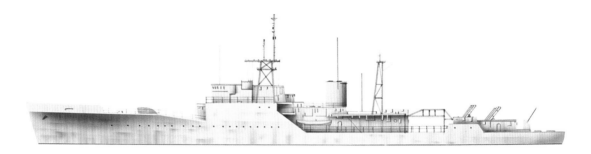

Largely prefabricated, the 12 frigates of *Grafton*'s class were too lightly gunned to be effective as escort ships. Their anti-submarine weapons consisted of two Limbo three-barrelled depth-charge launchers firing a pattern of large depth charges with great accuracy over a wide area. *Grafton* was broken up in 1971.

SPECIFICATIONS

Type:	British frigate
Displacement:	1480 tonnes (1456 tons)
Dimensions:	94.5m x 10m x 4.7m (310ft x 33ft x 15ft 6in)
Machinery:	Single screw, turbines
Top speed:	27.8 knots
Main armament:	Two 40mm (1.6in) guns
Launched:	September 1954

Centauro

Centauro was one of a class of four vessels built with US funds and equipped with automatic anti-submarine and medium anti-aircraft armament. The guns were mounted one above the other in the twin turrets, but this arrangement was later changed to conventional placing. *Centauro* was stricken in 1984.

SPECIFICATIONS

Type:	Italian frigate
Displacement:	2255 tonnes (2220 tons)
Dimensions:	104m x 11m x 4m (339ft x 38ft x 11ft 6in)
Machinery:	Twin screws, geared turbines
Main armament:	Four 76mm (3in) guns
Launched:	April 1954

TIMELINE

 1954    1955

Cigno

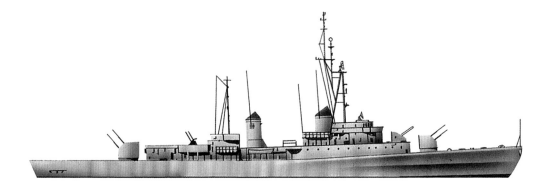

Of the same class as *Centauro* (see oppsite) but of greater displacement, *Cigno* shared the same features. The Italian-made 76mm (3in) guns were mounted in twin turrets and could fire 60 rounds per minute. In the 1960s, the turrets were replaced by three single 76mm (3in) mounts. *Cigno* was broken up in 1983.

SPECIFICATIONS	
Type:	Italian frigate
Displacement:	2455 tonnes (2220 tons)
Dimensions:	103m x 12m x 4m (339ft 3in x 38ft x 11ft 6in)
Machinery:	Twin screws, geared turbines
Main armament:	Four 76mm (3in) guns
Launched:	March 1955

Gatineau

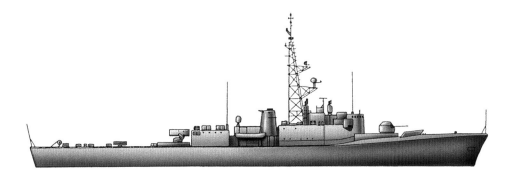

Developed from the *St Laurent* class, *Gatineau* was among four of its class to be modernized between 1966 and 1973, with variable-depth sonar and an ASROC launcher replacing an anti-submarine mortar and one 76mm (3in) gun. It served during the Gulf War. Decommissioned in 1996, *Gatineau* was broken up in 2009.

SPECIFICATIONS	
Type:	Canadian frigate
Displacement:	2641.6 tonnes (2600 tons)
Dimensions:	111.6m x 12.8m x 4.2m (366ft x 42ft x 13ft 2in)
Machinery:	Twin screws, geared turbines
Top speed:	28 knots
Main armament:	One Harpoon octuple SSM, two 76mm (3in) guns, one Mk15 Phalanx 20mm (0.79in) CIWS, six 324mm (12.75in) torpedo tubes
Complement:	290
Launched:	1957

Gemlik

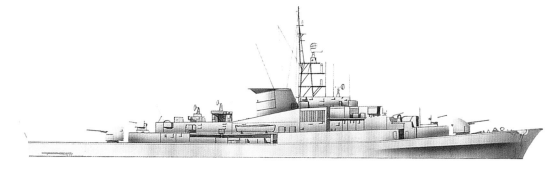

Emden, a *Köln*-class West German frigate, was transferred to the Turkish Navy in September 1983. It carried anti-submarine weapons, sensors and electronic counter-measures. Four launch tubes fired acoustic homing torpedoes. It could also lay up to 80 mines. After a fire, *Emden* was scrapped in 1994.

SPECIFICATIONS	
Type:	Turkish frigate
Displacement:	2743 tonnes (2700 tons)
Dimensions:	109.9m x 11m x 5.1m (360ft 7in x 36ft x 16ft 9in)
Machinery:	Twin screws, gas turbines/diesel engines
Top speed:	28 knots
Main armament:	Two 100mm (3.9in) guns, four 533mm (21in) torpedo tubes
Launched:	March 1959

1957    1959

Frigates of the 1960s & '70s: Pt1

Frigates, by the 1960s, were multi-task ships that fulfilled the role of the former light cruiser, able to provide a sufficiently powerful naval presence, whether helpful or punitive, in the event of some maritime incident or localized trouble-spot. Their missile armament equipped them for anti-aircraft or anti-submarine action, or both.

Carlo Bergamini

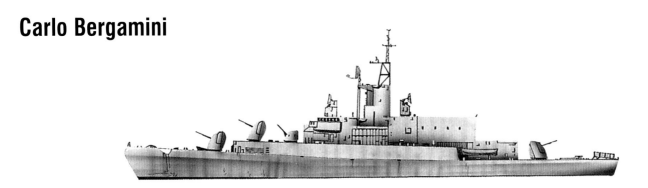

As well as fully automatic 76mm (3in) guns, this small but effective frigate carried a new type of single-barrelled mortar automatic depth charge discharger capable of firing 15 rounds per minute to a range of 920m (1000yd); two types of 304mm (12in) torpedo tube; and a helicopter. It was broken up in 1981.

SPECIFICATIONS

Type:	Italian frigate
Displacement:	1676 tonnes (1650 tons)
Dimensions:	94m x 11m x 3m (308ft 3in x 37ft 3in x 10ft 6in)
Machinery:	Twin screws, diesel motors
Main armament:	Three 76mm (3in) guns
Launched:	June 1960

Dido

Third of the *Leander* class frigates, *Dido* was one of eight to receive the GWS 40 Ikara ASW missile system, from 1978. It also carried a Wasp (later replaced by a Lynx) light helicopter. In 1983, *Dido* was sold to New Zealand to become HMNZS *Southland*. It was stricken in 1995 and broken up in India.

SPECIFICATIONS

Type:	British frigate
Displacement:	2844 tonnes (2800 tons)
Dimensions:	113m x 12m x 5.4m (372ft x 41ft x 18ft)
Machinery:	Twin screws, turbines
Top speed:	30 knots
Main armament:	Two 114mm (4.5in) guns, one quad launcher for Seacat missiles
Complement:	263
Launched:	December 1961

TIMELINE

 1960   1961   

Doudart de Lagrée

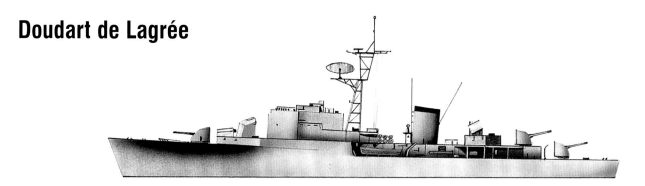

Doudart de Lagrée, intended for escort work and colonial patrol, could carry a force of 80 commandos. In the late 1970s, one gun turret was replaced by four Exocet missile launchers. Developments in building techniques and equipment design meant that the class was superseded, and the ship was stricken in 1991.

SPECIFICATIONS	
Type:	French frigate
Displacement:	2235 tonnes (2200 tons)
Dimensions:	102m x 11.5m x 3.8m (334ft x 37ft 6in x 12ft 6in)
Machinery:	Twin screws, diesels
Top speed:	25 knots
Main armament:	Three 100mm (3.9in) guns, twin anti-aircraft weapons
Complement:	210
Launched:	April 1961

Galatea

Galatea was an improvement of the Type 12 Rothesay class frigate. The missile system was housed on the extended superstructure forward of the bridge. The class was updated in the 1970s and 1980s. After serving in the Far East and Persian Gulf, Galatea was decommissioned in 1987 and sunk as a target in 1988.

SPECIFICATIONS	
Type:	British frigate
Displacement:	2906 tonnes (2860 tons)
Dimensions:	113.4m x 12.5m x 4.5m (372ft x 41ft x 14ft 9in)
Machinery:	Twin screws, turbines
Top speed:	28 knots
Main armament:	One anti-submarine Ikara missile launcher
Launched:	May 1963

Yubari

The Japanese Navy's Ishikari class was too small to carry the equipment, weapon stocks and electronic equipment vital for a late twentieth-century escort vessel, and the Yubari class is an enlarged version, with more fuel capacity and greater cruising range. A Phalanx CIWS was intended but not fitted.

SPECIFICATIONS		
Type:	Japanese frigate	
Displacement:	1777 tonnes (1690 tons)	
Dimensions:	91m x 10.8m x 3.5m (298ft 6in x 35ft 5in x 11ft 6in)	
Machinery:	Twin screws, gas turbines and diesels	
Top speed:	25 knots	
Main armament:	Two quad Harpoon SSM launchers, one 76mm (3in) gun, one 375mm (14.75in) mortar, six 324mm (12.75in) torpedo tubes	
Complement:	98	
Launched:	1982	

1963  1982

Frigates of the 1960s & '70s: Part 2

Distinctions between destroyers and frigates were increasingly blurred. Both types now routinely carried a helicopter, to aid in reconnoitring, anti-submarine attacks, and search-and-rescue missions. Many frigates were now equipped for anti-submarine warfare, with sonar detection and ASROC missile launchers.

Davidson

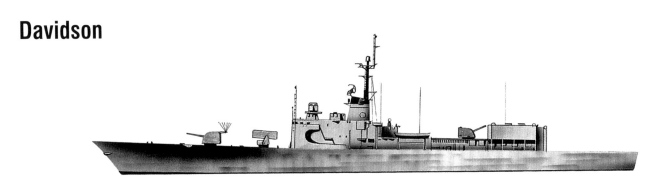

A *Garcia*-class destroyer escort, re-rated as a frigate from 1975, *Davidson* had gyro-driven stabilizers, enabling it to operate in heavy seas. A large box launcher held eight Asroc anti-submarine missiles. Twin torpedo tubes were later removed. Sold to Brazil in 1989 as *Paraibo,* it was decommissioned in 2002.

SPECIFICATIONS

Type:	US frigate
Displacement:	3454 tonnes (3400 tons)
Dimensions:	126m x 13.5m x 7m (414ft 8in x 44ft 3in x 24ft)
Machinery:	Single screw, turbines
Top speed:	27 knots
Main armament:	Two 127mm (5in) dual-purpose guns
Complement:	270
Launched:	1964

Carabiniere

Supplementary gas turbines gave *Carabiniere* extra speed when needed. Mast and funnel were an integrated structure. Anti-submarine weapons were a single semi-automatic depth charge mortar and six torpedo tubes, and two helicopters. Anti-missile defence was provided by SCLAR rockets. It was withdrawn in 2008.

SPECIFICATIONS

Type:	Italian frigate
Displacement:	2743 tonnes (2700 tons)
Dimensions:	113m x 13m x 4m (371ft x 43ft 6in x 12ft 7in)
Machinery:	Twin screws, diesels, gas turbines
Top speed:	20 knots (diesel), 28 knots (diesel and turbines)
Main armament:	Six 76mm (3in) guns
Launched:	September 1967

TIMELINE

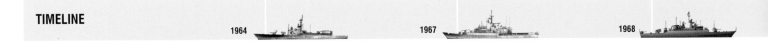

1964   1967   1968

Alvand

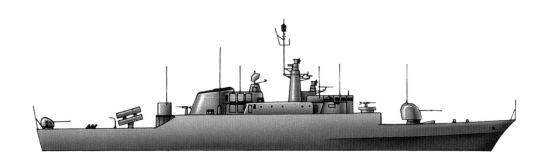

SPECIFICATIONS	
Type:	Iranian frigate
Displacement:	1564.7 tonnes (1540 tons)
Dimensions:	94.5m x 10.5m x 3.5m (310ft x 34ft 5in x 11ft 6in)
Machinery:	Twin screws, gas turbines and diesels
Top speed:	40 knots
Main armament:	One SSM launcher, one Seacat SAM system, one 114mm (4.5in) gun, one Limbo Mk10 anti-submarine mortar
Complement:	135
Launched:	1968

Known as the *Saam* class until the lead ship's name changed to *Alvand* in 1985, these craft were a British Vosper Thornycroft design, in effect a scaled-down version of the Type 21 frigate. The Seacat launcher was later removed. One ship of this class was sunk by US aircraft in 1988, but *Alvand* remains in service.

Downes

SPECIFICATIONS	
Type:	US frigate
Displacement:	4165 tonnes (4100 tons)
Dimensions:	126.6m x 14m x 7.5m (415ft 4in x 46ft 9in x 24ft 7in)
Machinery:	Single screw, turbines
Top speed:	28 + knots
Main armament:	One 127mm (5in) gun, one eight-tube Sea Sparrow missile launcher plus 20mm (0.79in) Phalanx CIWS
Launched:	December 1969

A large class (46 in total), these frigates were criticized for limited manoeuvrability and low anti-submarine capability. The midships tower structure was intended to carry an advanced electrical array, but this was not developed; *Downes* carried standard sea and air search radars instead. It was sunk as as a target in 2003.

Chikugo

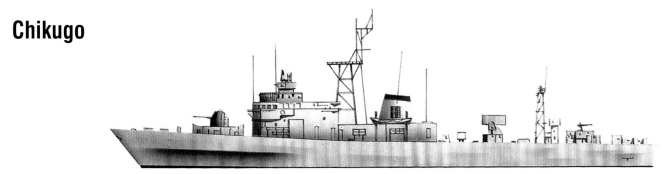

SPECIFICATIONS	
Type:	Japanese frigate
Displacement:	1493 tonnes (1470 tons)
Dimensions:	93m x 11m x 4m (305ft 5in x 35ft 5in x 11ft 6in)
Machinery:	Twin screws, diesels
Main armament:	Two 76mm (3in) guns
Launched:	1970

Chikugo was one of 11 units laid down in 1968. These frigates were the smallest ships to carry the anti-submarine weapon ASROC. Light anti-aircraft armament was installed, as the class were intended for inshore patrolling, protected by land-based fighter aircraft and missiles. *Chikugo* was decommissioned in 1996.

1969    1970

Frigates of the 1970s

While the functions of frigates and destroyers were tending to merge in deep-sea operations, a clear role for smaller versions of the frigate remained in coastal patrol work. Here there was a case for continued employment of the forward gun, in work that requires stop-and-search techniques and the interception of fast-moving craft.

Athabaskan

Athabaskan and three sister-ships were designed for anti-submarine warfare. Two hangars housed Sea King helicopters, giving the ships more flexibility than other anti-submarine vessels of the period. The armament now includes a SAM launcher, a Mk15 20mm (0.79in) Phalanx CIWS, and six 324mm (12.75in) torpedo tubes.

SPECIFICATIONS	
Type:	Canadian frigate
Displacement:	4267 tonnes (4200 tons)
Dimensions:	129.8m x 15.5m x 4.5m (426ft x 51ft x 15ft)
Machinery:	Twin screws, gas turbines
Top speed:	30 knots
Main armament:	One 127mm (5in) gun, one triple mortar
Complement:	285
Launched:	1970

Izumrud

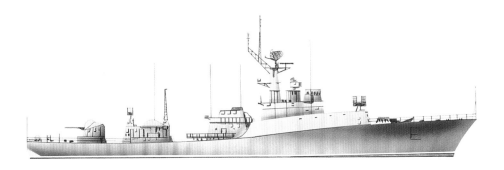

Used by the KGB Border Guard on inshore patrol, *Izumrud* carried twin 533mm (21in) torpedo tubes, SAM SA-N-4 missiles and rocket launchers. Turbines developed 24,000hp; the diesels produced 16,000hp. Range was 1805km (950 miles) at 27 knots, 8550km (4500 miles) at 10 knots. Disposal details are unknown.

SPECIFICATIONS	
Type:	Soviet frigate
Displacement:	1219 tonnes (1200 tons)
Dimensions:	72m x 10m x 3.7m (236ft 3in x 32ft 10in x 12ft 2in)
Machinery:	Triple screws, one gas turbine, two diesel engines
Main armament:	Two 57mm (2.24in) guns, SAM missiles
Complement:	310
Launched:	1970

TIMELINE

1970   1972

Najin

Equipped with Russian guns of World War II vintage and SS-N-2A missiles removed from redundant Soviet vessels, this was one of two similar ships built in North Korea in the early 1970s. Their later history is unclear. The names may have been changed and it is likely that both have been withdrawn from service.

SPECIFICATIONS	
Type:	North Korean frigate
Displacement:	1524 tonnes (1500 tons)
Dimensions:	100m x 9.9m x 2.7m (328ft x 32ft 6in x 8ft 10in)
Machinery:	Twin screws, diesels
Top speed:	33 knots
Main armament:	Two 100mm (3in) guns, three 533mm (21in) torpedo tubes
Complement:	180
Launched:	1972

Baptista de Andrade

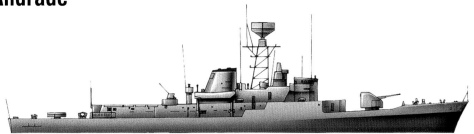

The four frigates of this class were ill-armed by contemporary standards, being deficient in anti-aircraft and anti-submarine defences. Portugal hoped to sell the quartet to Colombia in 1977, but the deal did not materialize. The *Andrades* are used only as coastal patrol vessels and are not deployed with NATO ships.

SPECIFICATIONS	
Type:	Portuguese frigate
Displacement:	1423.4 tonnes (1401 tons)
Dimensions:	84.6m x 10.3m x 3.3m (277ft 8in x 33ft 10in x 10ft 10in)
Machinery:	Twin screws, diesels
Top speed:	24.4 knots
Main armament:	One 100mm (3.9in) gun, six 324mm (12.75in) torpedo tubes
Complement:	113
Launched:	1973

Broadsword

Broadsword was the first general-purpose frigate designed to follow the *Leander* class. It was planned to build 26 units armed with missiles only, the main anti-submarine weapon being the Lynx helicopter. Later groups were fitted with extra weapons and sensors. *Broadsword* was sold to Brazil as *Greenhalgh* in 1995.

SPECIFICATIONS	
Type:	British frigate
Displacement:	4470 tonnes (4400 tons)
Dimensions:	131m x 15m x 4m (430ft 5in x 48ft 8in x 14ft)
Machinery:	Twin screws, gas turbines
Main armament:	Four M38 Exocet launchers, two 40mm (1.6in) guns
Complement:	407
Launched:	1975

1973   1975

Frigates of the 1970s & '80s

Most navies saw the frigate as the core-vessel of the surface fleet, capable of most tasks. Britain, for example, had 55 destroyers and 84 frigates in 1960. Twenty years later, it had 13 destroyers but still retained 53 frigates. The trend was very much towards fewer, but more versatile and comprehensively armed, warships.

Lupo

The *Lupo* class is an effective design, used by the Italian and other navies. Integral to the operating of *Lupo* was the SADOC automated combat control system, which enabled it to work with similarly fitted ships in an integrated group. Two helicopters are carried. *Lupo* was sold to Peru as *Palacios* in 2005.

SPECIFICATIONS	
Type:	Italian frigate
Displacement:	2540 tonnes (2500 tons)
Dimensions:	112.8m x 12m x 3.6m (370ft 2in x 39ft 4in x 12ft)
Machinery:	Twin screws, gas turbines plus diesels
Top speed:	35 knots
Main armament:	Eight Otomat SSM, one Sea Sparrow SAM launcher, one 127mm (5in) gun, six 324mm (12.75in) torpedo tubes
Complement:	185
Launched:	1976

Mourad Rais

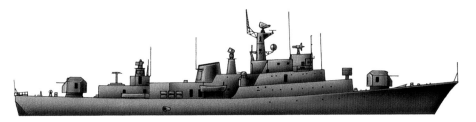

Mourad Rais was Soviet-built and equipped, of the 'Koni' class of light frigate intended for export to nations allied with or friendly to Soviet Russia. Three were supplied to Algeria between 1978 and 1984, intended mainly for anti-submarine use. A Russian-managed modernization programme was under way in 2009–10.

SPECIFICATIONS	
Type:	Algerian frigate
Displacement:	1930.4 tonnes (1900 tons)
Dimensions:	95m x 12.8m x 4.2m (311ft 8in x 42ft 13ft 9in)
Machinery:	Triple screws, diesels plus gas turbine
Top speed:	27 knots
Main armament:	One twin SA-N-4 SAM launcher, four 76mm (3in) guns, two RBU-6000 anti-submarine rocket launchers, two depth-charge racks
Complement:	110
Launched:	1978

TIMELINE

1976   1978   1980

Godavari

SPECIFICATIONS	
Type:	Indian frigate
Displacement:	4064 tonnes (4000 tons)
Dimensions:	126.5m x 14.5m x 9m (415ft x 47ft 7in x 29ft 6in)
Machinery:	Twin screws, turbines
Top speed:	27 knots
Main armament:	Two 57mm (2.24in) guns, four SS-N-2C Styx missiles, SA-N-4 Gecko missiles
Complement:	313
Launched:	May 1980

Godavari is a modified British *Leander*-class frigate, but with Russian and Indian weapon systems as well. Two Sea King or Chetak helicopters can be housed in the hangar. The ship is an early example of the trend towards a smooth profile to minimize visibility on radar. Most of the armament is mounted on the foredeck.

Admiral Petre Barbuneanu

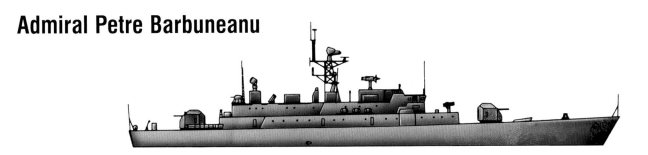

SPECIFICATIONS	
Type:	Romanian frigate
Displacement:	1463 tonnes (1440 tons)
Dimensions:	95.4m x 11.7m x 3m (303ft 1in x 38ft 4in x 9ft 8in)
Machinery:	Twin screws, diesels
Top speed:	24 knots
Main armament:	Four 76mm (3in) guns, two rocket launchers, four 533mm (21in) torpedo tubes
Complement:	95
Launched:	1981

Romania's navy operates in the Black Sea, and this vessel was intended for anti-submarine work and fitted with 16-tube RBU-2500 anti-submarine mortars. With three similar ships, it forms the 'Tetal' class. Sensor equipment included hull-mounted sonar, air/surface search radar and fire control radar.

Doyle

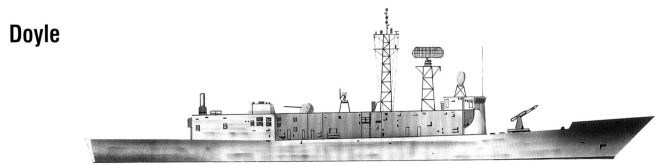

SPECIFICATIONS	
Type:	US frigate
Displacement:	3708 tonnes (3650 tons)
Dimensions:	135.6m x 14m x 7.5m (444ft 10in x 45ft x 24ft 7in)
Machinery:	Single screw, gas turbines
Top speed:	28 knots
Main armament:	One 76mm (3in) gun, Harpoon missile launcher
Launched:	May 1982

Doyle's profile is unlike that of previous frigates, an almost complete departure from earlier post-World War II designs. It reveals a warship reliant on missiles (not guns), built to have a minimal radar image, and with sophisticated radar and sonar detection systems. Its two helicopters carry anti-submarine weapons.

1981  1982

Frigates of the 1980s

Advances in missile technology made ships of frigate type less dependent on the traditional 127mm (5in) guns, and in the 1980s they often carried only a single main gun. Some even dispensed with that, though guns are useful for firing salutes. CIWS systems, based on 20mm (0.79in) or 30mm (1.18in) gun combinations, were introduced.

Jacob van Heemskerck

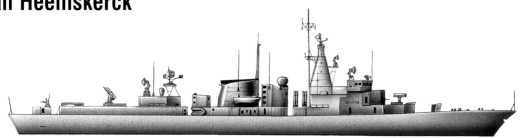

The Chilean *Admiral Latorre* since 2005, this Dutch L-class missile frigate was completed in 1986 for air defence. Dispensing with the usual forward gun, it has only two 20mm guns. Surface-search, air/surface search and fire control radars are fitted, along with hull-mounted sonar. Its sister-ship was also sold to Chile.

SPECIFICATIONS

Type:	Dutch frigate
Displacement:	3810 tonnes (3750 tons)
Dimensions:	130.2m x 14.4m x 6m (427ft x 47ft x 20ft)
Machinery:	Twin screws, gas turbines
Top speed:	30 knots
Main armament:	Eight Harpoon SSM, Standard SM-1MR SAM, Sea Sparrrow octuple launcher, Goalkeeper 30mm (1.18in) CIWS, four 324mm (12.75in) torpedo tubes
Complement:	197
Launched:	1983

Al Madina

Designed and built in France, *Al Madina* is one of four frigates supplied to Saudi Arabia in the mid-1980s and intended for general-purpose use, but chiefly ship-to-ship fighting. They can operate a SA 365F Dauphin helicopter, though it is not regularly carried. A full suite of radar equipment is fitted, and hull-mounted sonar.

SPECIFICATIONS

Type:	Saudi Arabian frigate
Displacement:	2651.8 tonnes (2610 tons)
Dimensions:	115m x 12.5m x 4.9m (377ft 3in x 41ft x 16ft)
Machinery:	Twin screws, diesels
Top speed:	30 knots
Main armament:	Eight Otomat Mk 2 SSM, one Crotale SAM launcher, one 100mm (3.9in) gun, four 440mm (17.33in) torpedo tubes
Complement:	179
Launched:	1983

TIMELINE

1983 1984

Kotor

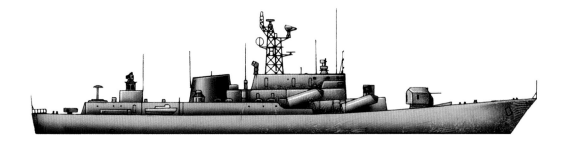

SPECIFICATIONS	
Type:	Yugoslav frigate
Displacement:	1930.4 tonnes (1900 tons)
Dimensions:	96.7m x 12.8m x 4.2m (317ft 3in x 42ft x 13ft 9in)
Machinery:	Triple screws, diesels plus gas turbine
Top speed:	27 knots
Main armament:	Four SS-N-2C SSM, one twin SA-N-4 SAM launcher, two 76mm (3in) guns, six 324mm (12.75in) torpedo tubes, two RBU-6000 anti-submarine rocket launchers
Complement:	110
Launched:	1984

Built on the general plan of the Soviet 'Koni'-class frigates, *Kotor* and *Pula* were larger, with various structural variations. Armament and mechanical equipment were Russian and Western, reflecting the non-aligned staus of Yugoslavia (as it then was). *Kotor* is now part of Montenegro's navy, but not operational.

Jianghu III

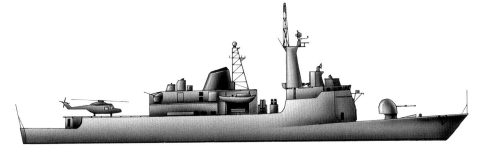

SPECIFICATIONS	
Type:	Chinese frigate
Displacement:	1895 tonnes (1865 tons)
Dimensions:	103.2m x 10.83m x 3.1m (338ft 7in x 35ft 6in x 10ft 2in)
Machinery:	Twin screws, diesels
Top speed:	25.5 knots
Main armament:	Eight UJ-1 Eagle Strike SSM, four 100mm (3.9in) guns, two anti-submarine mortars, two depth-charge racks
Complement:	180
Launched:	1986

Following on from 'Jianghu I and II', this class incorporates more up-to-date anti-ship weaponry and a roomier superstructure. Low-powered, they are intended for coastal anti-submarine patrol work rather than deep-sea duty. Many boats have been sold to other navies, including Pakistan, Egypt and Thailand.

Inhaúma

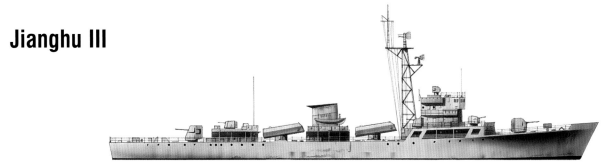

SPECIFICATIONS	
Type:	Brazilian frigate
Displacement:	2001.5 tonnes (1970 tons)
Dimensions:	95.8m x 11.4m x 5.5m (314ft 3in x 37ft 5in x 18ft)
Machinery:	Twin screws, diesels and gas turbine
Top speed:	27 knots
Main armament:	Four MM40 Exocet SSM, one 114mm (4.5in) gun, six 324mm (12.75in) torpedo tubes
Complement:	162
Launched:	1986

Inhaúma was intended as the first of 16 light patrol ships, forming a major element in the Brazilian Navy. Designed in Germany, they carry a range of equipment, including a Swedish fire-control system, British combat data system, American engines, and French missile systems. The flight deck takes a Lynx helicopter.

1986

Frigates of the 1980s & '90s

Stealth technology is standard in modern frigates. Superstructures and hulls are designed to offer a minimal radar cross section, with low profiles and large areas of smooth walling. This reduces air resistance, improving speed and manoeuvrability. At the same time, air and surface search radar has widened its detection capacity.

Halifax

SPECIFICATIONS	
Type:	Canadian frigate
Displacement:	4826 tonnes (4750 tons)
Dimensions:	134.1m x 16.4m x 4.9m (440ft 9in x 53ft 9in x 16ft 2in)
Machinery:	Twin screws, gas turbine and diesel
Top speed:	28 knots
Main armament:	Eight Harpoon SSM, two VLS for Sea Sparrow SAM, one 57mm (2.24in) gun, one Mk 15 Phalanx 20mm (0.79in) CIWS, four 324mm (12.75in) torpedo tubes
Complement:	225
Launched:	1988

Canada's 'City' class are big frigates with a large funnel offset to port. *Halifax* was completed in June 1992. The armament carried is chiefly for surface and aerial defence, but the helicopter that each ship carries is normally equipped for anti-submarine action. A comprehensive ship-by-ship class refit began in 2007.

Neustrashimyy

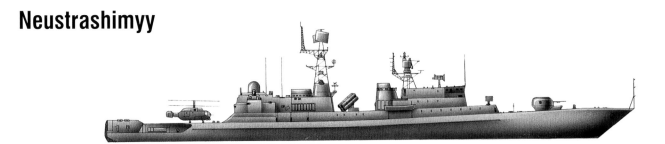

SPECIFICATIONS	
Type:	Soviet frigate
Displacement:	3556 tonnes (3500 tons)
Dimensions:	130m x 15.5m x 5.6m (426ft 6in x 50ft 11in x 18ft 5in)
Machinery:	Twin screws, gas turbines
Top speed:	32 knots
Main armament:	One SS-N-25 SSM launcher, one SA-N-9 SAM launcher, two CADS-N-1 gun/missile CIWS, one RBU-12000 anti-submarine rocket launcher
Complement:	210
Launched:	1988

Introduced to improve the Soviet Navy's anti-submarine capacities, this class has four ships. The flat-flared hull, divided superstructure and funnel shape reduce and disperse the ship's radar returns. Its own sensors include towed sonar. In 2008–09, *Neustrashimyy* was deployed to the Somali coast to help combat piracy.

TIMELINE

1988   1989

Thetis

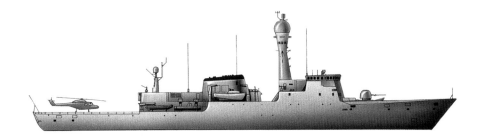

The *Thetis* class of four ships was intended to strengthen Denmark's fleet in the late twentieth and early twenty-first centuries. It was intended to mount Harpoon and Sea Sparrow missiles, but this plan was abandoned at the end of the Cold War and the class serve as lightly armed, outsize fishery protection ships.

SPECIFICATIONS	
Type:	Danish patrol ship
Displacement:	3556 tonnes (3500 tons)
Dimensions:	112.5m x 14.4m x 6m (369ft 1in x 47ft 3in x 19ft 8in)
Machinery:	Single screw, diesels
Top speed:	21.5 knots
Main armament:	One 76mm (3in) gun, depth-charge racks
Complement:	61
Launched:	1989

Floréal

Described as 'surveillance frigates' to protect France's 'exclusive economic zone', this class has been built using mercantile ship techniques and modular assembly, rather than the construction typical of warships. It can embark a helicopter up to Super Puma size. Two ships of the class were built for Morocco.

SPECIFICATIONS	
Type:	French frigate
Displacement:	2997 tonnes (2950 tons)
Dimensions:	93.5m x 14m x 4.3m (307ft x 46ft x 14ft)
Machinery:	Twin screws, diesels
Top speed:	20 knots
Main armament:	Two MM38 Exocet SSM, one 100mm (3.9in) gun
Complement:	80 plus 24 armed troops
Launched:	1990

Nareusan

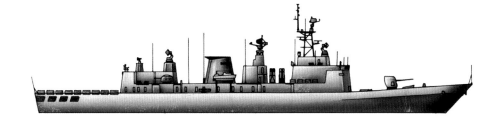

Built in China, to the design of the 'Jianghu' class, *Nareusan* was fitted out in Thailand, using Western machinery, weapons and electronic systems that make it more effective than the 'Jianghus'. Air/surface radar, navigation radar, fire-control radar and hull-mounted sonar are fitted, and it can embark a Lynx-type helicopter.

SPECIFICATIONS	
Type:	Thai frigate
Displacement:	3027.7 tonnes (2980 tons)
Dimensions:	120m x 13m x 3.81m (393ft 8in x 42ft 8in x 12ft 6in)
Machinery:	Twin screws, gas turbines and diesels
Top speed:	32 knots
Main armament:	Eight Harpoon SSM, one Mk 41 VLS for Sea Sparrow SAM, one 127mm (5in) gun, six 324mm (12.75in) torpedo tubes
Complement:	150
Launched:	1993

1990    1993

Minehunters & Sweepers

In the second half of the twentieth century, mines became more varied in type and in the methods of laying. New generation 'smart' mines could be programmed in various ways to sense and counter the activity of minehunters. Several American ships were damaged by mines during operations in the Persian Gulf region.

Edera

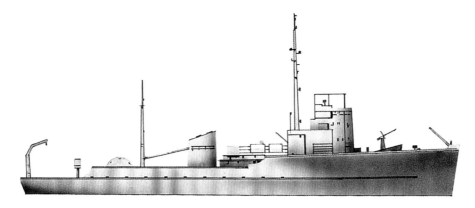

SPECIFICATIONS

Type:	Italian minesweeper
Displacement:	411 tonnes (405 tons)
Dimensions:	44m x 8m x 2.6m (144ft 26ft 6in x 8ft 6in)
Machinery:	Twin screws, diesel engine
Top speed:	14 knots
Main armament:	Two 20mm (0.8in) anti-aircraft guns
Complement:	38
Launched:	1955

Edera was one of the 19-strong *Agave* class of minesweepers, of non-magnetic wood and alloy composite construction, and designed for inshore minesweeping duties. During the 1960s, the class was part of Italy's countermining force. Fuel carried was 25 tonnes (25 tons), enough for 4750km (2500 miles) at 10 knots.

Bambú

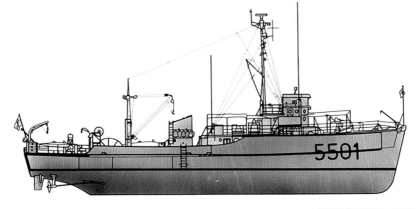

SPECIFICATIONS

Type:	Italian coastal minesweeper
Displacement:	375 tonnes (370 tons)
Dimensions:	44.1m x 8.5m x 2.6m (144ft 5in x 28ft 8ft 6in)
Machinery:	Twin screws, diesel engine
Top speed:	13 knots
Main armament:	Two 20mm (0.79in) anti-aircraft guns
Complement:	31
Launched:	1956

One of four converted American *Adjutant*-class minesweepers, *Bambú* entered service in 1956. Equipped with radar and sonar, it was wooden-hulled to defeat magnetic mines. Ships of this class were often assigned to United Nations' coastal patrol work in troubled regions. They were phased out in the 1990s.

TIMELINE

1955   1956   1957

Dromia

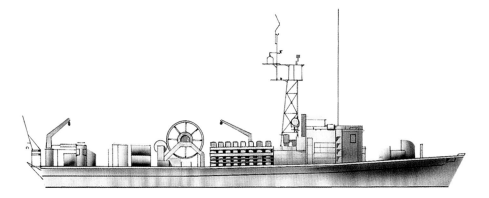

Dromia was one of 20 inshore minesweepers of the British 'Ham' class built in Italy between 1955 and 1957. They were designed to operate in shallow waters, rivers and estuaries, and when first built they were a new type of minesweeper, embodying many of the lessons learned during World War II and later hostilities.

SPECIFICATIONS	
Type:	Italian minesweeper
Displacement:	132 tonnes (130 tons)
Dimensions:	32m x 6.4m x 1.8m (106ft x 21ft x 6ft)
Machinery:	Twin screws, diesel engine
Top speed:	14 knots
Main armament:	One 20mm (0.79in) gun
Launched:	1957

Eridan

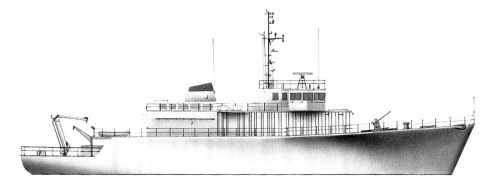

In the late 1970s, France, Belgium and the Netherlands combined to build 35 minehunters to a design that could be adapted by each nation. *Eridan* could be used for minehunting, minelaying, extended patrols, training, directing unmanned minesweeping craft, and as an HQ ship for diving operations.

SPECIFICATIONS	
Type:	French minehunter
Displacement:	552 tonnes (544 tons)
Dimensions:	49m x 8.9m x 2.5m (161ft x 29ft 2in x 8ft 2in)
Machinery:	Single screw, diesel engine
Top speed:	15 knots
Main armament:	One 20mm (0.79in) gun
Launched:	February 1979

Aster

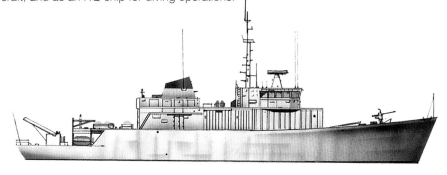

Aster is a Belgian example of the Tripartite minehunter design designed for NATO service. France, Belgium and the Netherlands each built its own hulls, which were fitted out in Belgium with French electronics and Dutch machinery. All vessels carry full nuclear, biological and chemical (NBC) protection and minesweeping equipment, and can be used as patrol and surveillance craft.

SPECIFICATIONS	
Type:	Belgian minehunter
Displacement:	605 tonnes (595 tons)
Dimensions:	51.5m x 8.9m x 2.5m (169ft x 29ft x 8ft)
Machinery:	Single screw, diesel engine, two manoeuvring propellers and one bowthruster
Top speed:	15 knots
Main armament:	One 20mm (0.79in) anti-aircraft gun
Launched:	1981

1979   1981

Specialized Naval Ships

As in previous decades, fleet support required a variety of dedicated craft. The main difference was that the complexity of modern operating systems made it imperative to build ships for purpose rather than to adapt existing ones, though there were exceptions. Other specialisms included assault and landing ships.

Filicudi

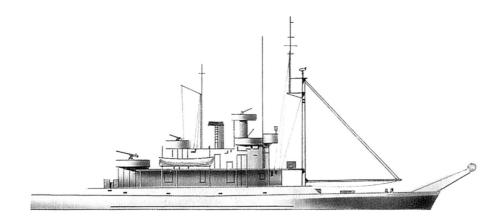

Filicudi and its sister *Alicudi* were based on a standard NATO design, and could lay nets of various depths across harbour entrances. A large, open deck for handling the nets is situated in the low bow section. A boom attached to the foremast controls the lifting and lowering of nets.

SPECIFICATIONS

Type:	Italian net layer
Displacement:	847 tonnes (834 tons)
Dimensions:	50m x 10m x 3.2m (165ft 4in x 33ft 6in x 10ft 6in)
Machinery:	Twin screws, diesel-electric motors
Top speed:	12 knots
Main armament:	One 40mm (1.57in) gun
Launched:	September 1954

Caorle

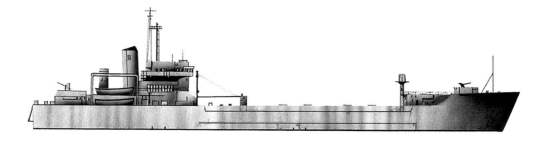

In 1972, USS *York County* was sold to Italy, and renamed *Caorle*. It could carry up to 575 fully equipped assault troops, or a mixture of troops, tanks and other vehicles. A flat bottom and shallow draught allowed the bow to be grounded on a shallow beach for unloading. It was scrapped at Naples in 1999.

SPECIFICATIONS

Type:	Italian landing ship
Displacement:	8128 tonnes (8000 tons)
Dimensions:	135m x 19m x 5m (444ft x 62ft x 16ft 6in)
Machinery:	Twin screws, diesel engines
Top speed:	17.5 knots
Main armament:	Six 76mm (3in) guns
Launched:	March 1957

TIMELINE   1954 1957  1959

Chazhma

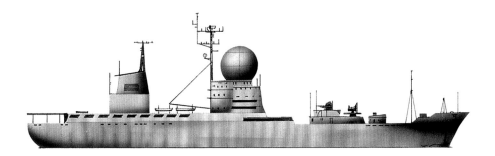

A 7381-tonne (7265-ton) bulk ore carrier of the *Dshankoy* class, *Chazhma* was converted into a missile range ship in 1963 to serve in the Pacific. A 'Ship Globe' radar was mounted in the dome above the bridge. A helicopter platform and hangar, built into the aft superstructure, let it operate one 'Hormone' helicopter.

SPECIFICATIONS	
Type:	Soviet missile range ship
Displacement:	13,716 tonnes (13,500 tons)
Dimensions:	140m x 18m x 8m (458ft x 59ft x 26ft)
Machinery:	Twin screws, diesel engines
Top speed:	15 knots
Launched:	1959

Deutschland

The first West German ship to exceed the post-war limit of 3048 tonnes (3000 tons), *Deutschland* carried a range of armaments for training purposes, including 100mm (3.9in) and 40mm (1.57in) guns, depth-charge launchers, mines and torpedoes. It was towed to India for scrapping in 1994.

SPECIFICATIONS	
Type:	German training ship
Displacement:	5588 tonnes (5500 tons)
Dimensions:	145m x 18m x 4.5m (475ft 9in x 59ft x 14ft 9in)
Machinery:	Triple screws, diesel motors, turbines
Top speed:	22 knots
Main armament:	Four 100mm (3.9in) guns
Complement:	500 including 267 cadets
Launched:	1960

Alligator class

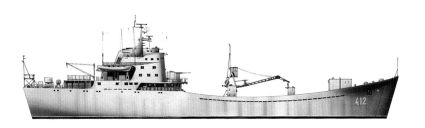

The Project 1171 Nosorog large landing craft were designated 'Alligator' by NATO. Sixteen were built, with bow and stern ramps. All carried some weapons and at least one crane. Up to 30 armoured personnel carriers and their troops could be transported. Some were used in the South Ossetia War of 2008.

SPECIFICATIONS	
Type:	Soviet landing ship
Displacement:	4775 tonnes (4700 tons) full load
Dimensions:	112.8m x 15.3m x 4.4m (370ft 6in x 50ft 2in x 14ft 5in)
Machinery:	Twin screws, diesels
Top speed:	18 knots
Main armament:	Two/three SA-N-5 SAM launchers, one 122mm (4.8in) rocket launcher
Complement:	75 plus 300 combat troops
Launched:	1964

1960

1964

Post-War Conventional Submarines: Part 1

The German and Japanese submarine fleets had been eliminated by the end of World War II, but the Russians, Americans and British still had substantial numbers. Post-war building programmes began only in the 1950s, stimulated by the Cold War.

Whiskey

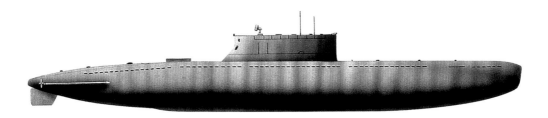

About 240 of these attack submarines were built between 1951 and 1958. Four units were converted to early warning boats between 1959 and 1963, but from 1963 the long-range Bear aircraft reduced their strategic importance in some areas. By the 1980s, these submarines had disappeared from the effective list.

SPECIFICATIONS

Type:	Soviet submarine
Displacement:	1066 tonnes (1050 tons) [surface], 1371 tonnes (1350 tons) [submerged]
Dimensions:	76m x 6.5m x 5m (249ft 4in x 21ft 4in x 16ft)
Machinery:	Twin screws, diesel engines [surface], electric motors [submerged]
Top speed:	18 knots [surface], 14 knots [submerged]
Main armament:	Two 406mm (16in), four 533mm (21in) torpedo tubes
Launched:	1956

Golf I

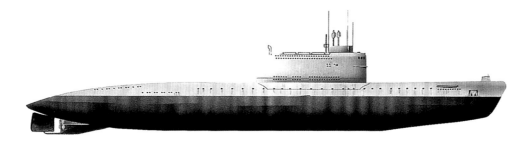

Twenty-three Golf I-class submarines were completed between 1958 and 1962, entering service at a rate of six to seven per year. The ballistic missiles were housed vertically in the rear section of the extended fin. Many boats in the class were modified after commissioning. All were withdrawn by 1990.

SPECIFICATIONS

Type:	Russian missile submarine
Displacement:	2336 tonnes (2300 tons) [surface], 2743 tonnes (2700 tons) [submerged]
Dimensions:	100m x 8.5m x 6.6m (328ft x 27ft 11 in x 21ft 8in)
Machinery:	Triple screws, diesel engine (surface), electric motors [submerged]
Top speed:	17 knots (surface), 12 knots [submerged]
Main armament:	Three SS-N-4 ballistic missiles, ten 533mm (21in) torpedo tubes
Launched:	1957

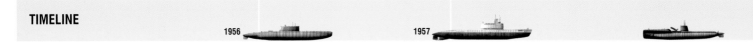

TIMELINE

1956   1957

Grayback

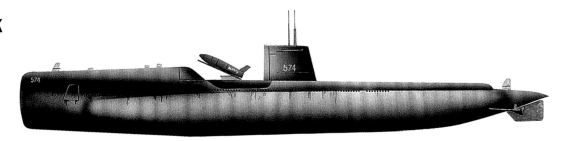

During construction, *Grayback* was altered to carry the first naval cruise missile, Regulus, and did so until 1964. Recommissioned in 1968 as an amphibious transport submarine for undercover missions, it carried 67 marines and their SEAL swimmer delivery vehicles. *Grayback* was sunk as a target in April 1986.

SPECIFICATIONS	
Type:	US submarine
Displacement:	2712 tonnes (2670 tons) [surface], 3708 tonnes (3650 tons) [submerged]
Dimensions:	102m x 9m (335ft x 30ft)
Machinery:	Twin screws, diesel engines [surface], electric motors [submerged]
Main armament:	Four Regulus missiles, eight 533mm (21in) torpedo tubes
Launched:	1957

Daphné

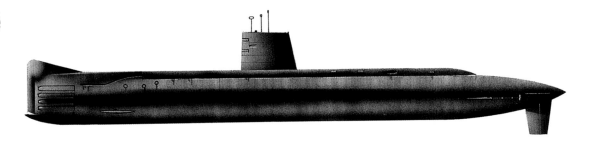

Eleven submarines of this class were launched between 1964 and 1970. The double hull had a deep keel to improve stability. They had good manoeuvrability, low noise, a small crew and were easy to maintain; several navies bought them. *Daphné* was decommissioned in 1989; the rest of the class followed by 1996.

SPECIFICATIONS	
Type:	French submarine
Displacement:	884 tonnes (870 tons) [surface], 1062 tonnes (1045 tons) [submerged]
Dimensions:	58m x 7m x 4.6m (189ft 8in x 22ft 4in x 15ft)
Machinery:	Twin screws, diesel [surface], electric drive [submerged]
Top speed:	13.5 knots [surface], 16 knots [submerged]
Main armament:	Twelve 552mm (21.7in) torpedo tubes
Launched:	June 1959

Dolfijn

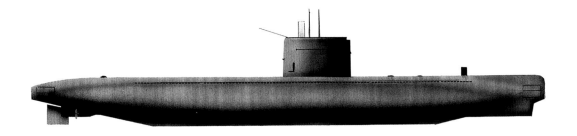

Dolfijn and three sister boats were built to a unique triple-hulled design – three cylinders arranged in a triangular shape. The upper cylinder housed the crew, navigational equipment and armament; the lower cylinders, the powerplant. Maximum diving depth was almost 304m (1000ft). *Dolfijn* was broken up in 1985.

SPECIFICATIONS	
Type:	Dutch submarine
Displacement:	1518 tonnes (1494 tons) [surface], 1855 tonnes (1820 tons) [submerged]
Dimensions:	80m x 8m x 4.8m (260ft 10in x 25ft 9in x15ft 9in)
Machinery:	Twin screws, diesel [surface], electric motors [submerged]
Main armament:	Eight 533mm (21in) torpedo tubes
Launched:	May 1959

1959

Post-War Conventional Submarines: Part 2

In the 1960s, Japan, Italy and West Germany resumed the building of submarines, placing their forces within the NATO and US-Japanese defence agreements. Several non-aligned nations, notably Sweden, also produced effective submarine types.

Enrico Toti

The four vessels of this class were the first submarines to be built in Italy since World War II. The design was revised several times before the hunter/killer model, for use in shallow and confined waters, was finally approved. Withdrawn in 1992, *Enrico Toti* is now a museum vessel in Milan.

SPECIFICATIONS

Type:	Italian submarine
Displacement:	532 tonnes (524 tons) [surface], 591 tonnes (582 tons) [submerged]
Dimensions:	46.2m x 4.7m x 4m (151ft 7in x 15ft 5in x 13ft)
Machinery:	Single screw, diesel engines [surface], electric motors [submerged]
Top speed:	14 knots surfaced, 15 knots submerged
Main armament:	Four 533mm (21in) torpedo tubes
Launched:	March 1967

Harushio

This was Japan's first post-World War II fleet submarine class, named for Oshio, though it had a different bow shape to its successors. The primary role, in Japan's Maritime Defence Force, was to act as targets in anti-submarine training exercises. *Harushio* was withdrawn and scrapped in 1984.

SPECIFICATIONS

Type:	Japanese submarine
Displacement:	1676.4 tonnes (1650 tons) surfaced, 2184.4 tonnes (2150 tons) [submerged]
Dimensions:	88m x 8.2m x 4.9m (288ft 8in x 26ft 11in x 16ft)
Machinery:	Twin screws, diesel [surface], electric motors [submerged]
Top speed:	18 knots surfaced, 14 knots submerged
Main armament:	Eight 533mm (21in) torpedo tubes
Complement:	80
Launched:	1967

TIMELINE

1967   1968

U-12

U-12 was one of the first class of German submarines built after World War II. They were a successful type, with over 40 boats serving in foreign navies. The hull was of a non-magnetic steel alloy. Diesel engines developed 2300hp, and the single electric motor developed 1500hp. U-12 was decommissioned in 2005.

SPECIFICATIONS	
Type:	German submarine
Displacement:	425 tonnes (419 tons) [surface], 457 tonnes (450 tons) [submerged]
Dimensions:	43.9m x 4.6m x 4.3m (144ft x 15ft x 14ft)
Machinery:	Single screw, diesel engine [surface], electric motors [submerged]
Top speed:	10 knots [surface], 17 knots [submerged]
Main armament:	Eight 533mm (21in) torpedo tubes
Launched:	1968

Näcken

Näcken and its two sister craft were re-engined in 1987–88 with a closed-air independent propulsion system (AIP), which let them operate underwater for up to 14 days without surfacing. Withdrawn in the 1990s, they have been replaced by the Swedish-built Vastergötland class, which uses the same technology.

SPECIFICATIONS	
Type:	Swedish submarine
Displacement:	995.7 tonnes (980 tons) [surface], 1169 tonnes (1150 tons) [submerged]
Dimensions:	49.5m x 5.7m x 5.5m (162ft 5in x 18ft 8in x 18ft)
Machinery:	Single screw, diesel engine [surface], electric motors [submerged]
Top speed:	20 knots [surface], 25 knots [submerged]
Main armament:	Six 533mm (21in), two 400mm (15.75in) torpedo tubes
Complement:	19
Launched:	1978

Kilo

The 'Kilo' class were the first Soviet boats to use a modern teardrop hull form, which gives a good underwater speed-power ratio. Double-hulled, they are fast and highly manoeuvrable, well suited to operations in restricted waters. About 17 continue in service with Russia, and 33 in other fleets.

SPECIFICATIONS	
Type:	Russian submarine
Displacement:	2336 tonnes (2300 tons) [surface], 2946 tonnes (2900 tons) [submerged]
Dimensions:	73m x 10m x 6.5m (239ft 6in x 32ft 10in x 21ft 4in)
Machinery:	Single screw, diesel engine [surface], electric motor [submerged]
Top speed:	12 knots [surface], 18 knots [submerged]
Main armament:	Six 533mm (21in) torpedo tubes
Launched:	1981

1978  1981

Post-War Conventional Submarines: Part 3

While by the 1980s the nuclear submarine had taken up the major strategic role, its deployment was confined to five navies. A significant tactical part was still played by conventionally powered boats, particularly for coastal patrol.

Galerna

One of four medium-range submarines built in Spain to the design of the French *Agosta* class, *Galerna* marked a major step forward in Spanish submarine technology. The original weapon-stock was 16 reload torpedoes or nine torpedoes and 19 mines. State-of-the-art sonar kit is carried.

SPECIFICATIONS	
Type:	Spanish submarine
Displacement:	1473 tonnes (1450 tons) [surface], 1753 tonnes (1725 tons) [submerged]
Dimensions:	67.6m x 6.8m x 5.4m (221ft 9in x 22ft 4in x 17ft 9in)
Machinery:	Single screw, diesel engine [surface], electric motor [submerged]
Top speed:	12 knots surfaced, 20 knots submerged
Main armament:	Four 551mm (21.7in) torpedo tubes
Launched:	December 1981

Walrus

Walrus was not completed until 1991, due to a fire, and *Zeeleeuw*, commissioned in 1989, became class leader. The use of high-tensile steel gave a diving depth of 300m (985ft). New Gipsy fire control and electronic command systems reduced crew numbers to 49. A class refit in 2007 extended the service life of these boats.

SPECIFICATIONS	
Type:	Dutch submarine
Displacement:	2490 tonnes (2450 tons) [surface], 2845 tonnes (2800 tons) [submerged]
Dimensions:	67.5m x 8.4m x 6.6m (222ft 27ft 7in x 21ft 8in)
Machinery:	Single screw, diesel engines [surface], electric motors [submerged]
Top speed:	13 knots [surface], 20 knots [submerged]
Main armament:	Four 533mm (21in) torpedo tubes
Launched:	October 1985

TIMELINE

1981   1985   1986

Hai Lung

Hai Lung is a modified Dutch *Zwaardvis*-class, probably the most efficient design of the 1970s. They are quiet boats, with all machinery mounted on anti-vibration mountings. They carried up to 28 Tigerfish acoustic homing torpedoes, and in 2005 were upgraded to carry UGM-84 Harpoon anti-ship missiles.

SPECIFICATIONS	
Type:	Taiwanese submarine
Displacement:	2414 tonnes (2376 tons) [surface], 2702 tonnes (2660 tons) [submerged]
Dimensions:	66.9m x 8.4m x 6.7m (219ft 5in x 27ft 6in x 22ft)
Machinery:	Single screw, diesel engines [surface], electric motors [submerged]
Top speed:	11 knots [surface], 20 knots [submerged]
Main armament:	Six 533mm (21in) torpedo tubes
Complement:	67
Launched:	October 1986

Upholder

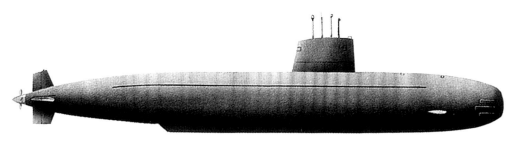

Designed as a new Royal Navy conventionally-powered patrol submarine, all four in the Upholder class were transferred to Canada in 1998. The teardrop-shaped hull is of high tensile steel, enabling dives to 200m (656ft). The class has had technical problems, and three were undergoing refits in 2009–10.

SPECIFICATIONS	
Type:	British submarine
Displacement:	2220 tonnes (2185 tons) [surface], 2494 tonnes (2455 tons) [submerged]
Dimensions:	70.3m x 7.6m x 5.5m (230ft 8in x 25ft x 18ft)
Machinery:	Single screw, diesel engine [surface], electric motors [submerged]
Top speed:	12 knots [surface], 20 knots [submerged]
Main armament:	Six 533mm (21in) torpedo tubes
Launched:	December 1986

Collins

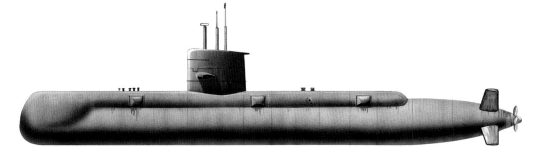

Swedish-designed and Australian-built, this attack submarine was marred by mechanical and electronic problems when introduced in 1995 and the combat data management system was replaced. Three of the six boats in the *Collins* class are active, the others in reserve. Plans for replacement were put in hand in 2007.

SPECIFICATIONS	
Type:	Australian submarine
Displacement:	2220 tonnes (3051 tons) [surface], 2494 tonnes (3353 tons) [submerged]
Dimensions:	77.5m x 7.8m x 7m (254ft x 25ft 7in x 23ft)
Machinery:	Single screw, diesel-electric engines [surface], electric motor [submerged]
Top speed:	10 knots [surface], 20 knots [submerged]
Main armament:	Six 533mm (21in) torpedo tubes
Launched:	1993

1993

Nuclear Submarines: Part 1

Until 1954, the submarine was essentially a surface ship that could submerge. Experiments had been going on with power sources that would allow for indefinite submergence. A hydrogen peroxide motor was tried in the 1940s, but nuclear reaction offered the answer: in 1954 USS *Nautilus* became the first 'true' submarine.

Nautilus

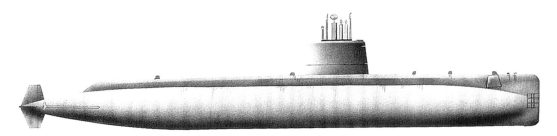

The world's first nuclear-powered submarine, *Nautilus* was of conventional design. Early trials established many records, including nearly 2250km (1400 miles) submerged in 90 hours at 20 knots, and a passage beneath the ice over the North Pole. Stricken in 1980, *Nautilus* is preserved at Groton, Connecticut.

SPECIFICATIONS

Type:	US submarine
Displacement:	4157 tonnes (4091 tons) [surface], 4104 tonnes (4040) [submerged]
Dimensions:	98.7m x 8.4m x 6.6m (323ft 9in x 27ft 8in x 21ft 9in)
Machinery:	Twin screws, nuclear reactor, turbines
Top speed:	23 knots [submerged]
Main armament:	Six 533mm (21in) torpedo tubes
Complement:	105
Launched:	January 1954

Skipjack

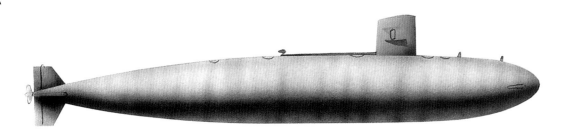

With a teardrop-form hull, and diving planes on the fin, *Skipjack* was fast and manoeuvrable. No stern tubes were fitted: the aft hull shape tapered sharply. It introduced the S5W fast-attack propulsion plant used in all subsequent attack and ballistic submarines until the *Los Angeles*. The class was withdrawn by the 1990s.

SPECIFICATIONS

Type:	US submarine
Displacement:	3124 tonnes (3075 tons) [surface], 3570 tonnes (3513 tons) [submerged]
Dimensions:	76.7m x 9.6m x 8.9m (251ft 8in x 31ft 6in x 29ft 2in)
Machinery:	Single screw, nuclear reactor, turbines
Top speed:	18 knots surfaced, 30 knots submerged
Main armament:	Six 533mm (21in) torpedo tubes
Complement:	93
Launched:	May 1958

TIMELINE   1954   1958   1959

USS Halibut

First deployed with cruise missiles, *Halibut* was used for secret intelligence work, often involving the retrieval of objects of military interest from the sea-bed. Midget submarines, carried in the former Regulus missile space, were used for this work. *Halibut* was decommissioned in 1976, and broken up in 1994.

SPECIFICATIONS	
Type:	US submarine
Displacement:	(3846 tons) surfaced, (4895 tons) submerged
Dimensions:	106.7m x 9m x 6.3m (350ft x 29ft 6in x 20ft 9in)
Machinery:	Twin screws, one nuclear reactor
Top speed:	15 knots [surfaced], 15.5 knots [submerged]
Main armament:	Five SSM -N-8 Regulus 1 or two SSM-N-9 Regulus II, six 533mm (21in) torpedo tubes
Complement:	111
Launched:	1959

George Washington

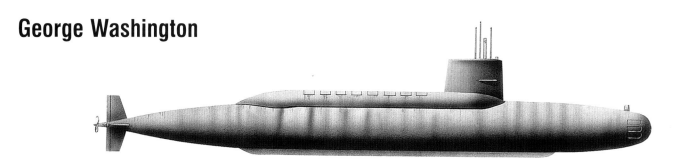

In 1955, the Soviet Union began modifying submarines to carry nuclear-tipped ballistic missiles. At the time, the United States was developing the Polaris A1 missile, and the submarine *Scorpion* was adapted to carry it. Renamed *George Washington,* it was 'de-missiled' in the 1980s, and decommissioned in 1986.

SPECIFICATIONS	
Type:	US ballistic missile submarine
Displacement:	6115 tonnes (6019 tons) [surface], 6998 tonnes (6888 tons) [submerged]
Dimensions:	116.3m x 10m x 8.8m (381ft 7in x 33ft x 28ft 10in)
Machinery:	Single screw, one pressurized water-cooled reactor, turbines
Top speed:	20 knots [surface], 30.5 knots [submerged]
Main armament:	Sixteen Polaris missiles, six 533mm (21in) torpedo tubes
Launched:	June 1959

Dreadnought

Britain's first nuclear-powered submarine, *Dreadnought* was a detect-and-destroy vessel. The form of the hull was based on the shape of a whale. Its power-plant was an American S5W reactor as fitted to the US *Skipjack* submarines. Laid up in 1982, *Dreadnought* was towed to Rosyth for disposal in the following year.

SPECIFICATIONS	
Type:	British submarine
Displacement:	3556 tonnes (3500 tons) [surface], 4064 tonnes (4000 tons) [submerged]
Dimensions:	81m x 9.8m x 8m (265ft 9in x 32ft 3in x 26ft)
Machinery:	Single screw, nuclear reactor, steam turbines
Top speed:	20 knots [surface], 30 knots [submerged]
Main armament:	Six 533mm (21in) torpedo tubes
Complement:	88
Launched:	October 1960

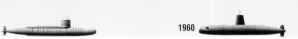

1960

Resolution

Lead boat of a class of four, designed to carry Britain's nuclear deterrent armament of US Polaris missiles, *Resolution* went on its first patrol in 1968. It was intended that one of the four submarines would always be on active service. With the deployment of the *Vanguard* class with Trident missiles, *Resolution* was retired in 1994.

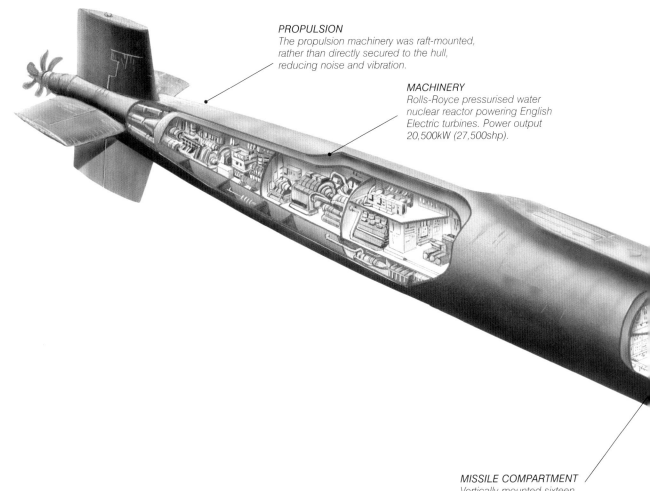

PROPULSION
The propulsion machinery was raft-mounted, rather than directly secured to the hull, reducing noise and vibration.

MACHINERY
Rolls-Royce pressurised water nuclear reactor powering English Electric turbines. Power output 20,500kW (27,500shp).

MISSILE COMPARTMENT
Vertically-mounted sixteen Polaris A3 missiles, in two rows of eight. These had a range of 4,631km (2,500 nautical miles) and multiple nuclear warheads.

Resolution

The *Resolution* class submarines were armed with the American Polaris SLBM and they took over the British deterrent role from the RAF in 1968. Their characteristics were very similar to the American *Lafayettes*.

SPECIFICATIONS	
Type:	British submarine
Displacement:	7620 tonnes (7500 tons) [surface], 8636 tonnes (8500 tons) [submerged]
Dimensions:	129.5m x 10m x 9.1m (425ft x 33ft x 30ft)
Machinery:	Single screw, pressurized water reactor, geared steam turbines
Top speed:	20 knots [surfaced], 25 knots [submerged]
Main armament:	Sixteen UGM-27C Polaris A-3 SLBM, six 533mm (21in) torpedo tubes
Complement:	143
Launched:	September 1966

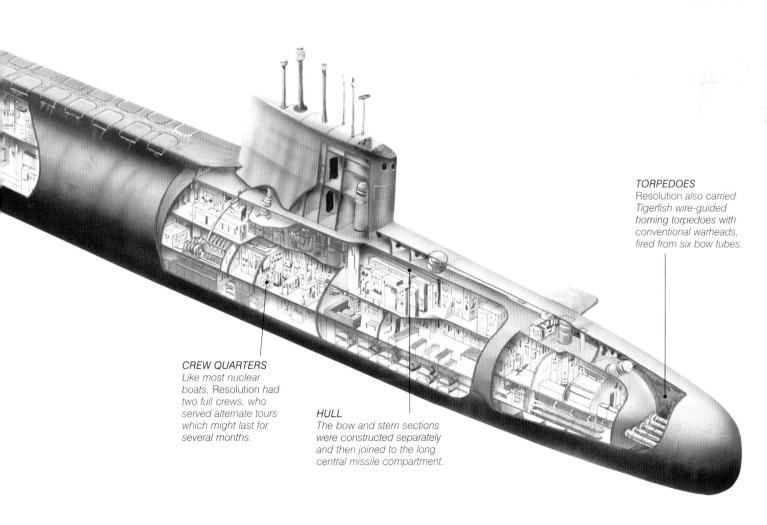

CONVERSION
The four Resolution *class boats were adapted in the mid-1980s to carry new Polaris AT-K missiles with British Chevaline multiple re-entry warheads.*

TORPEDOES
Resolution *also carried Tigerfish wire-guided homing torpedoes with conventional warheads, fired from six bow tubes.*

CREW QUARTERS
Like most nuclear boats, Resolution had two full crews, who served alternate tours which might last for several months.

HULL
The bow and stern sections were constructed separately and then joined to the long central missile compartment.

Nuclear Submarines: Part 2

Two US nuclear submarines, *Thresher* (1963) and *Scorpion* (1968), were lost in the 1960s. The loss of *Thresher* prompted the introduction of the Deep Sea Rescue Vessel (DSRV). Nuclear boats first carried conventional torpedoes but developments in rocket technology led to the introduction of the missile-firing submarine.

Daniel Boone

Daniel Boone was one of a sub-class of the *Lafayette* strategic missile nuclear submarines. Completed in April 1964, and fitted to carry UGM 73A Poseidon missiles, it was one of 12 boats adapted for the more reliable Trident type in 1980. With the advent of the *Ohio* class, *Daniel Boone* was retired in 1994.

SPECIFICATIONS

Type:	US submarine
Displacement:	7366 tonnes (7250 tons) [surface], 8382 tonnes (8250 tons) [submerged]
Dimensions:	130m x 10m x 10m (425ft x 33ft x 33ft)
Machinery:	Single screw, single water-cooled nuclear reactor, turbines
Top speed:	20 knots [surface], 35 knots [submerged]
Main armament:	Sixteen Polaris missiles, four 533mm (21in) torpedo tubes
Launched:	June 1962

Warspite

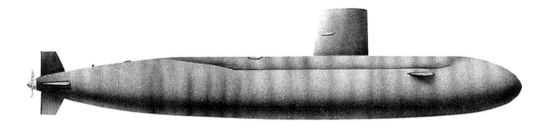

Using old battleship names confirmed nuclear submarines as the new capital ships. One of five boats in Britain's first class of nuclear submarines, *Warspite* also had an emergency battery, diesel generator and electric motor. It was withdrawn in 1991, when hairline cracks were found in its primary coolant circuit.

SPECIFICATIONS

Type:	British submarine
Displacement:	4368 tonnes (4300 tons) [surface], 4876 tonnes (4800 tons) [submerged]
Dimensions:	87m x 10m x 8.4m (285ft x 33ft 2in x 27ft 7in)
Machinery:	Single screw, pressurized water-cooled nuclear reactor, turbines
Top speed:	28 knots [submerged]
Main armament:	Six 533mm (21in) torpedo tubes
Launched:	1965

TIMELINE

1962   1965

George Washington Carver

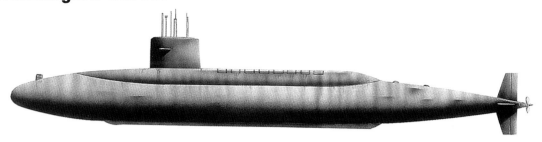

One of 29 vessels in the *Lafayette* class, *George Washington Carver* could dive to depths of 300m (985ft), and the nuclear core provided enough energy to propel the boat for 760,000km (400,000 miles). Its missile tubes were deactivated in 1991 and it became an attack boat. It was stricken and sent for recycling in 1993.

SPECIFICATIONS	
Type:	US submarine
Displacement:	7366 tonnes (7250 tons) [surface], 8382 tonnes (8250 tons) [submerged]
Dimensions:	129.5m x 10m x 9.6m (424ft 10in x 33ft 2in x 31ft 6in)
Machinery:	Single screw, one pressurized water-cooled nuclear reactor
Speed:	20 knots [surface], 30 knots [submerged]
Main armament:	Sixteen Trident C4 missiles, four 533mm (21in) torpedo tubes
Launched:	August 1965

Narwhal

The United States' 100th nuclear submarine, *Narwhal* was an attack boat, of the *Sturgeon* class. These were larger than the *Thresher* and *Permit* class, powered by an improved reactor. *Narwhal* was notably quiet and much of its activity was in reconnaissance, or eavesdropping. It was stricken in 1999.

SPECIFICATIONS	
Type:	US submarine
Displacement:	4374 tonnes (4246 tons) [surface], 4853.4 tonnes (4777 tons) [submerged]
Dimensions:	89.1m x 9.6m x 7.8m (292ft 3in x 31ft 8in x 25ft 6in)
Machinery:	Single screw, one pressurized water-cooled nuclear reactor
Speed:	26 knots [submerged]
Main armament:	Four 533mm (21in) torpedo tubes
Launched:	August 1965

Yankee

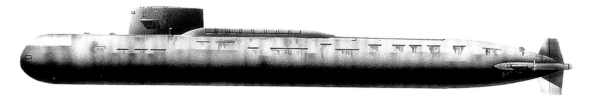

Project 667A, known to NATO as the 'Yankee' class, were more powerful than previous Russian nuclear submarines. Thirty-four boats were built, capable of firing missiles from underwater, and patrolled the US eastern seaboard. By the SALT arms limitation agreement, all strategic 'Yankees' were withdrawn by 1994.

SPECIFICATIONS	
Type:	Soviet submarine
Displacement:	7823 tonnes (7700 tons) [surfaced], 9450 tonnes [9300 tons] [submerged]
Dimensions:	132m x 11.6m x 8m (433ft 10in x 38ft 1in x 26ft 4in)
Machinery:	Twin screws, nuclear reactors, turbines
Top speed:	13 knots [surfaced], 27 knots [submerged]
Main armament:	Sixteen SS-N-6 missile tubes, six 533mm (21in) torpedo tubes
Complement:	120
Launched:	1967

1967

Nuclear Submarines: Part 3

France launched its first class of nuclear submarines armed with ballistic missiles with *Le Redoutable* in 1967, and China produced the 'Han' class in 1972. The invisibility of the nuclear submarine and the secrecy surrounding its cruising missions, together with its destructive power, added to the tensions of the Cold War.

Charlie II

Following from the 12 smaller 'Charlie I' (NATO code) as the Soviet Navy's Project 670M, the 'Charlie II' class carried SS-N-9 Siren anti-ship missiles, which could be fitted with nuclear warheads, and two sizes of torpedo. The class shadowed US carrier battle groups. Six were built, serving until the mid-1990s.

SPECIFICATIONS

Type:	Soviet missile submarine
Displacement:	4368.8 tonnes (4300 tons) [surface], 5181.6 tonnes (5100 tons) [submerged]
Dimensions:	103.6m x 10m x 8m (340 x 32ft 10in x 28ft)
Machinery:	Single screw, nuclear reactor
Top speed:	24 knots surfaced
Main armament:	Eight SS-N-9 cruise missiles, four 533mm (21in) and four 406mm (16in) torpedo tubes
Complement:	98
Launched:	1967

Delta I

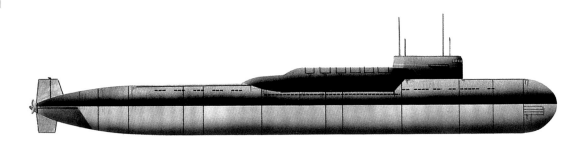

Between 1972 and 1977, Russia moved ahead in the Cold War with Project 667B – 18 large 'Delta'-class vessels, armed with new missiles that could out-range the US Poseidons. Initial tests showed the SS-N-48 missiles had a range of over 7600km (4000 miles). The class was scrapped between 1995 and 2004.

SPECIFICATIONS

Type:	Russian submarine
Displacement:	11,176 tonnes (11,000 tons) submerged
Dimensions:	150m x 12m x 10.2m (492ft x 39ft 4in x 33ft 6in)
Machinery:	Twin screws, two nuclear reactors, turbines
Top speed:	19 knots [surface], 25 knots [submerged]
Main armament:	Twelve SS-N-48 missile tubes, six 457mm (18in) torpedo tubes
Launched:	1971

TIMELINE

1967 1971 1972

Han

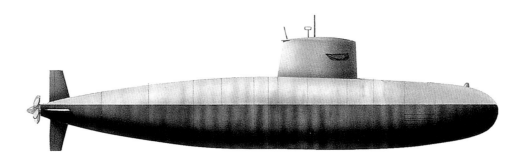

The Chinese Navy went nuclear in the early 1970s with the 'Han'-class attack submarines. The highly streamlined hull shape, based upon the US vessel *Albacore*, is a departure from previous Chinese submarine designs. Five boats were completed, of which two or three were considered still operational in 2010.

SPECIFICATIONS	
Type:	Chinese submarine
Displacement:	5080 tonnes (5000 tons) [submerged], surface displacement tonnage unknown
Dimensions:	90m x 8m x 8.2m (295ft 3in x 26ft 3in x 27ft)
Machinery:	Single screw, one pressurized-water nuclear reactor with turbine drive
Top speed:	25 knots [submerged]
Main armament:	Six 533mm (21in) torpedo tubes
Launched:	1972

Los Angeles

Los Angeles was lead boat in the world's most numerous class of nuclear submarines, with 45 still serving in 2010. Later members of the class (built up to 1996) have been modified. Though intended as hunter-killer boats, they are capable of land attack with Tomahawk missiles, shown in Iraq and Afghanistan.

SPECIFICATIONS	
Type:	US submarine
Displacement:	6096 tonnes (6000 tons) [surface], 7010.4 tonnes (6900 tons) [submerged]
Dimensions:	109.7m x 10m x 9.8m (360ft x 33ft x 32ft 4in)
Machinery:	Single screw, pressurized-water nuclear reactor, turbines
Top speed:	31 knots [submerged]
Main armament:	Four 533mm (21in) torpedo tubes, up to eight Tomahawk cruise missiles
Complement:	127
Launched:	April 1976

Ohio

The *Ohio* class were the largest submarines built in the West, surpassed only by the Soviet 'Typhoon' class. The size was determined by the size of the reactor plant. *Ohio* was commissioned in 1981. Originally the class carried the Trident C-4 missile, but from the ninth boat they were built to carry the D-5 version.

SPECIFICATIONS	
Type:	US submarine
Displacement:	16,360 tonnes (16,764 tons) [surface], 19,050 tonnes (18,750 tons) [submerged]
Dimensions:	170.7m x 12.8m x 11m (560ft x 42ft x 36ft 5in)
Machinery:	Single screw, pressurized-water nuclear reactor, turbines
Top speed:	28 knots [surface], 30+ knots [submerged]
Main armament:	24 Trident missiles, four 533mm (21in) torpedo tubes
Launched:	April 1979

1976   1979

Nuclear Submarines: Part 4

Improved reactors encouraged construction of large nuclear submarines. Boats of the United State's *Ohio* class were giants, but dwarfed by the Russian 'Typhoon' boats. Later types were smaller, on grounds of efficiency as well as cost (it was estimated that the United States had spent $700,000,000 on development by 1998).

Victor III

Russia's nuclear-powered hunter-killer submarines were codenamed 'Victor' by NATO. 'Victor III' could fire the SS-N-16 missile, which delivers a conventional homing torpedo to a greater range than otherwise possible. At least 43 'Victors' were launched, 26 of them being 'Victor IIIs', with some still operational in 2010.

SPECIFICATIONS

Type:	Soviet submarine
Displacement:	6400 tonnes (6300 tons) [submerged]
Dimensions:	104m x 10m x 7m (347ft 8in x 32ft 10in x 23ft)
Machinery:	Single screw, pressurized water-cooled nuclear reactor, turbines
Top speed:	30 knots
Main armament:	Six 533mm (21in) torpedo tubes
Launched:	1978

Typhoon

'Typhoon' is the largest submarine yet built, nearly half as big again as the US *Ohio* class. The missile tubes are situated in two rows in front of the fin. It can force a way up through ice up to 3m (9ft 10in) thick. *Dmitry Donskoi,* of this class, was test-firing new Bulava-M missiles in 2008–09.

SPECIFICATIONS

Type:	Russian submarine
Displacement:	25,400 tonnes (24,994 tons) [surface], 26,924 tonnes (26,500 tons) [submerged]
Dimensions:	170m x 24m x 12.5m (562ft 6in x 78ft 8in x 41ft)
Machinery:	Twin screws, pressurized water-cooled nuclear reactors, turbines
Top speed:	27 knots [submerged]
Main armament:	Twenty SS-N-20 nuclear ballistic missiles, two 533mm (21in) and four 650mm (25.6in) torpedo tubes
Launched:	1979

TIMELINE

1978

1979

San Francisco

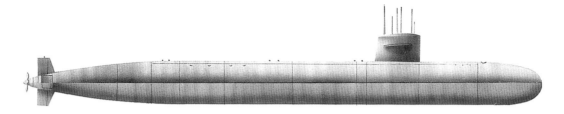

A *Los Angeles* class large attack submarine, *San Francisco* ran at full speed into an uncharted undersea mountain beneath the Pacific Ocean in January 2005, damaging the bows, but it managed to surface safely. Repaired with parts of withdrawn members of the class, it was restored to the fleet in 2008.

SPECIFICATIONS

Type:	US submarine
Displacement:	6300 tonnes (6200 tons) [surface], 7010 tonnes (6900 tons) [submerged]
Dimensions:	110m x 10m x 9.8m (360ft x 33ft x 32ft 4in)
Machinery:	Single screw, nuclear powered pressurized-water reactor, turbines
Top speed:	30+ knots [submerged]
Main armament:	Four 533mm (21in) torpedo tubes. Harpoon and Tomahawk missiles
Launched:	October 1979

Oscar I

Project 949 was for a class of submarines combining the 'Typhoon' class's Arktika reactor with the 'Victor III' sonar systems, able to launch cruise missiles while submerged and to spend up to 50 days under water. The missiles were placed between the inner and outer pressure hulls, at an angle of 40° from the vertical.

SPECIFICATIONS

Type:	Soviet submarine
Displacement:	tonnes (12,500 tons) [surface], 14,122 tonnes (13,900 tons) [submerged]
Dimensions:	143m x 18.21m x 8.99m (469ft 2in x 59ft 9in x 29ft 6in)
Machinery:	Twin screws, two pressurized water-cooled nuclear reactors
Top speed:	23 knots [submerged]
Main armament:	Four 533mm (21in) and four 650mm (25.6in) torpedo tubes launching SS-N-19, SS-N-15 and SS-N-16 Stallion missiles
Launched:	April 1981

Xia

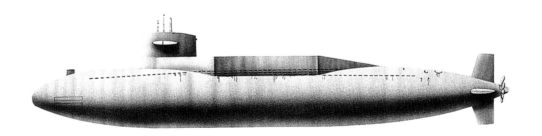

Type 092, 'Xia' was laid down in 1978. China's first ballistic missile submarine, it was an experimental craft. Two were built, of which one remains in service. The missiles were two-stage solid fuel rockets with inertial guidance for ballistic flight to 8000km (5000 miles), fitted with a nuclear warhead of two megatons.

SPECIFICATIONS

Type:	Chinese submarine
Displacement:	8128 tonnes (8000 tons), submerged
Dimensions:	120m x 10m x 8m (393ft 8in x 32ft 10in x 26ft 3in)
Machinery:	Single screw, pressurized water-cooled nuclear reactor
Top speed:	22 knots [submerged]
Main armament:	Twelve tubes for CSS-N-3 missiles, six 533mm (21in) torpedo tubes
Launched:	April 1981

1981  1981

Nuclear Submarines: Part 5

In the Cold War, nuclear submarines shadowed enemy fleet groups, eavesdropping on naval exercises. Doubtless the practice continues, but operational boats fall into two categories: strategic missile submarines armed with long-range nuclear weapons, and 'hunter-killer' submarines to intercept and destroy enemy vessels.

Georgia

An *Ohio*-class boat, *Georgia* was redesignated SSGN (guided missile) when modified to carry cruise missiles in 2004. A major refit and overhaul followed in 2008. The fin is set far forward, ahead of the missile tubes. Its nuclear reactor is shielded from the engine, control centre and living quarters.

SPECIFICATIONS

Type:	US submarine
Displacement:	16,865 tonnes (16,600 tons) [surface], 19,000 tonnes (18,700 tons) [submerged]
Dimensions:	170.7m x 12.8m x 10.8m (560ft x 42ft x 35ft 5in)
Machinery:	Single screw, pressurized water-cooled nuclear reactor, turbines
Top speed:	28 knots [surface], 30+ knots [submerged]
Main armament:	Twenty-four Trident missiles (C4), four 533mm (21in) torpedo tubes
Launched:	November 1982

Sierra

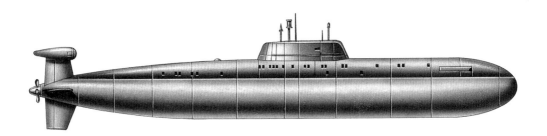

This was Project 945 and four boats were built before production switched to the 'Akula' boats of Project 971. Named 'Sierra' by NATO, the class used the Arktika reactor. Titanium-hulled, they had better safety provisions than previous Soviet nuclear submarines, including a crew escape pod. Three have been withdrawn.

SPECIFICATIONS

Type:	Russian submarine
Displacement:	7315 tonnes (7200 tons) [surface], 10,262 tonnes (10,100 tons) [submerged]
Dimensions:	107m x 12m x 8.8m (351ft x 39ft 5in x 28ft 11in)
Machinery:	Single screw, pressurized water-cooled nuclear reactor, turbines
Top speed:	8 knots [surfaced], 36 knots [submerged]
Main armament:	Four 533mm (21in) and four 650mm (25.6in) torpedo tubes with provision for SS-N-22 and SS-N-16 missiles
Complement:	61
Launched:	1983

TIMELINE

1982   1983   1985

Torbay

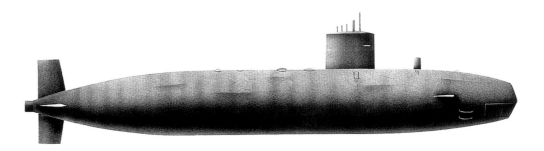

Torbay was one of the *Trafalgar* class of fleet submarine ordered in 1977, with a longer-life nuclear reactor. The main propulsion and auxiliary machinery raft are suspended from transverse bulkheads to maximize sound insulation. Anechoic tiles also reduce the acoustic signature. Modernizing of the class is taking place.

SPECIFICATIONS	
Type:	British submarine
Displacement:	4877 tonnes (4800 tons) [surface], 5384 tonnes (5300 tons) [submerged]
Dimensions:	85.4m x 10m x 8.2m (280ft 2in x 33ft 2in x 27ft)
Machinery:	Pump jet, pressurized water-cooled reactor, turbines
Main armament:	Five 533mm (21in) tubes for Tigerfish torpedoes
Launched:	March 1985

Vanguard

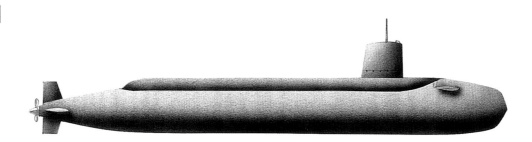

Vanguard carries 16 missiles in vertical launch tubes aft of the sail. Each can bear up to 14 warheads to targets more than 12,350km (6500 miles) distant. Like all submarines of this type, it operates independently, remaining submerged for months. The nuclear reactor is refitted and re-cored every eight years.

SPECIFICATIONS	
Type:	British submarine
Displacement:	15,240 tonnes (15,000 tons) [submerged]
Dimensions:	148m x 12.8m x 12m (486ft 6in x 42ft x 39ft 4in)
Machinery:	Single screw, pressurized water-cooled nuclear reactor
Top speed:	25+ knots [submerged]
Main armament:	Sixteen Trident D5 missiles, four 533mm (21in) torpedo tubes
Complement:	135
Launched:	1990

Le Triomphant

Le Triomphant is the first of France's 'new generation' missile submarines, commissioned in 1997 to replace the *Le Redoutable* class. It uses a new form of propeller. In February 2009, it and HMS *Vanguard* 'scraped' each other while on independent secret patrol beneath the Atlantic, but without consequences.

SPECIFICATIONS	
Type:	French submarine
Displacement:	12,842 tonnes (12,640 tons) [surface], 14,564.4 tonnes (14,335 tons) [submerged]
Dimensions:	138m x 17m x 12.5m (453ft x 55ft 8in x 41ft)
Machinery:	Single screw, nuclear reactor with pump jet propulsor
Top speed:	20 knots [surface] 25 knots [submerged]
Main armament:	M51 nuclear missiles from 2010
Launched:	1993

1990    1993

Kursk

Kursk was an 'Oscar II' class nuclear submarine, an attack boat with nuclear missiles. In 1999, it was deployed in the Mediterranean. It sank after internal explosions while on exercises with the Northern Fleet in the Barents Sea, on 12 August 2000. All the crew perished. A complex salvage operation retrieved the wreck in October 2001.

Kursk

Kursk sank down to 354ft (108m). Russia, Britain and Norway launched a rescue operation, but ten days after the explosions, the remaining crew was declared dead. Twenty-three out of the crew of 118 survived the explosions, but they were trapped in a compartment and died when their air ran out.

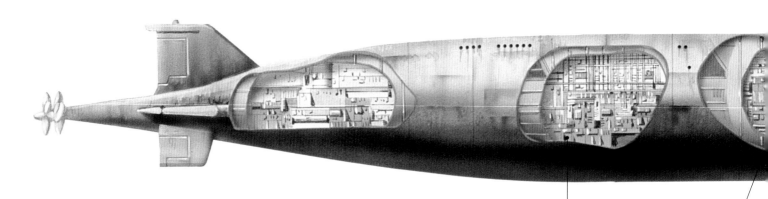

MACHINERY
Two pressurised water-cooled nuclear reactors powering two steam turbines. Power output 73,070kW (98,000shp).

FIFTH COMPARTMENT
This housed the nuclear reactors, and was protected by armoured steel walls 130mm (5.1in) thick, which withstood the blast.

SPECIFICATIONS

Type:	Russian submarine
Displacement:	14,834 tonnes (14,600 tons) [surface], 16,256 tonnes (16,000 tons) [submerged]
Dimensions:	154m x 18.21m x 8.99m (505ft 2in x 59ft 9in x 29ft 6in)
Machinery:	Twin screws, two pressurized water-cooled reactors powering steam turbines
Top speed:	16 knots [surface], 32 knots [submerged]
Main armament:	24 Granit cruise missiles, four 533mm (21in) and two 650mm (25.5in) torpedo tubes
Complement:	118
Launched:	1994

PROFILE
Like other Russian submarines, the 'Oscar II' class had an open bridge on the sail. The bulge probably housed an escape capsule.

EMERGENCY BUOY
This could have been automatically released to give a surface indication of a pressure-related problem, but it had been intentionally disabled.

BULKHEADS
Bulkheads separating the front compartments of the boat failed to prevent the blast effects from spreading back.

TORPEDO TUBES
A chemical explosion in tube No. 4 initiated the catastrophe. This was caused by the accidental combining of hydrogen peroxide and kerosene.

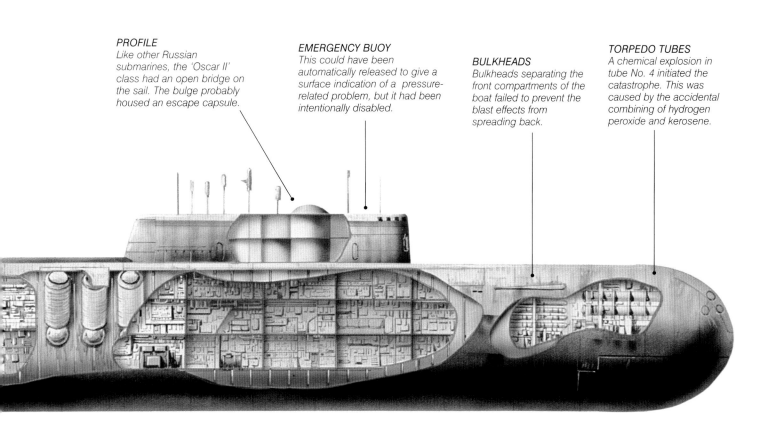

SECOND EXPLOSION
135 seconds after the first, a second larger explosion ripped open the third and fourth compartments.

Support and Repair Ships

The concept of the rapid-deployment task force – able to move at short notice to remote destinations, for military reasons or to provide post-disaster aid – depends on support and repair ships, which can keep up with other vessels in the force and provide the essentials of an operating base, often with command functions as well.

Hunley

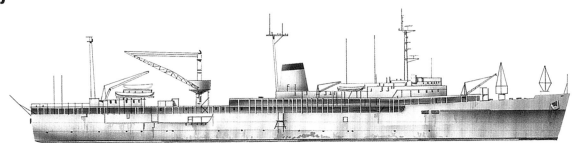

Hunley and its sister Holland were designed to provide repair and supply services to fleet ballistic missile submarines. With 52 workshops, Hunley could deal with the requirements of several submarines at once. It carried a helicopter for at-sea delivery. Decommissioned in 1994, it was broken up in 2007.

SPECIFICATIONS

Type:	US submarine tender
Displacement:	19,304 tonnes (19,000 tons)
Dimensions:	182.6m x 25.3m x 8.2m (599ft x 83ft x 27ft)
Machinery:	Single screw, diesel-electric engines
Main armament:	Four 20mm (0.79in) guns
Complement:	2490
Launched:	September 1961

Engadine

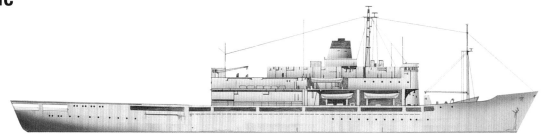

Laid down in August 1965, the Fleet Auxiliary Engadine was designed to train helicopter crews in deep-water operations. The large hangar aft of the funnel held four Wessex and two WASP helicopters, or two of the larger Sea Kings. Engadine could also operate pilotless target aircraft. It was scrapped in 1996.

SPECIFICATIONS

Type:	British helicopter support ship
Displacement:	9144 tonnes (9000 tons)
Dimensions:	129.3m x 17.8m x 6.7m (424ft 3in x 58ft 5in x 22ft)
Machinery:	Single screw, diesel engine
Top speed:	16 knots
Complement:	81 plus 113 training crew
Launched:	1966

TIMELINE

1961  1966 1970

Basento

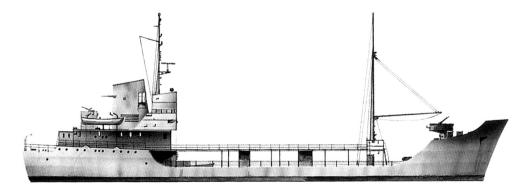

Basento was an auxiliary vessel supplying fresh water to the Italian fleet. Tank capacity is 1016 tonnes (1000 tons) and the ship's range is nearly 5700km (3000 miles) at 7 knots. The machinery space is placed aft. In 2009, Basento was given to Ecuador, to become a water supply ship for the arid Galapagos Islands.

SPECIFICATIONS	
Type:	Italian naval water tanker
Displacement:	1944 tonnes (1914 tons)
Dimensions:	66m x 10m x 4m (216ft 6in x 33ft x 13ft)
Machinery:	Twin screws, diesels
Top speed:	12.5 knots
Armament:	Two light anti-aircraft guns
Launched:	1970

Fort Grange

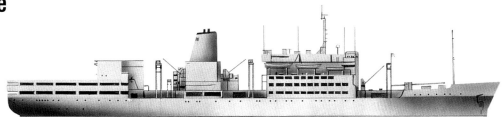

Up to 3500 tonnes of stores can be carried on board, to refuel and restock ships at sea in operational conditions. In addition. it can fly up to four Sea King helicopters on combat missions. *Fort Grange* spent 1994–2000 in the Adriatic Sea. Renamed *Fort Rosalie* in 2000, it was refitted in 2008–09.

SPECIFICATIONS	
Type:	British fleet replenishment ship
Displacement:	23,165 tonnes (22,800 tons)
Dimensions:	183.9m x 24.1m x 8.6m (603ft x 79ft x 28ft 2in)
Machinery:	Single screw, diesels
Top speed:	22 knots
Main armament:	Two 20mm (0.79in) cannon
Complement:	140 plus 36 aircrew
Launched:	1976

Frank Cable

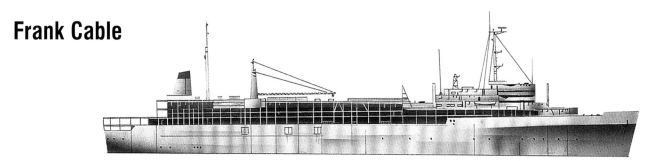

One of three improved versions of the *Spear* class, *Frank Cable* was equipped to support the *Los Angeles* class of submarine; up to four of this type can be handled simultaneously. These vessels are a far cry from World War II tenders, which were often old warships on their last role before scrapping.

SPECIFICATIONS	
Type:	US submarine tender
Displacement:	23,368 tonnes (23,000 tons)
Dimensions:	196.9m x 25.9m x 7.6m (646ft x 85ft x 25ft)
Machinery:	Single screw, turbines
Top speed:	18 knots
Main armament:	Two 40mm (1.57in) guns
Launched:	1978

1976  1978

Bulk Carriers

Large bulk carriers have played an important part in international economic life. The transport of huge tonnages has reduced the cost of shipping heavyweight raw materials like coal and iron ore. Automation speeds up loading and unloading at each end of the ships' run, and the ships are managed by a handful of crew.

Yeoman Burn

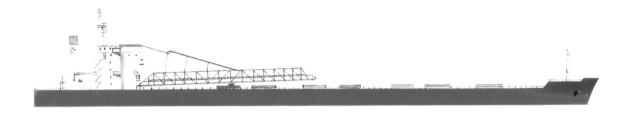

Of typical bulk carrier design, with high superstructure and engines at the stern, *Yeoman Burn*'s most notable feature is the lattice booms of the self-loading and unloading machinery. This enables the ship to use ports that lack automated loading. The ship is now under German ownership as *Bernhard Olendorff*.

SPECIFICATIONS

Type:	Norwegian bulk carrier
Displacement:	78,740 tonnes (77,500 tons)
Dimensions:	245m x 32.2m x 14m (830ft 10in x 105ft 8in x 46ft)
Machinery:	Single screw, diesel engines
Top speed:	14.6 knots
Cargo:	iron ore, coal, limestone, salt, coke or grain in bulk
Routes:	International routes
Complement:	25
Launched:	October 1990

Jakob Maersk

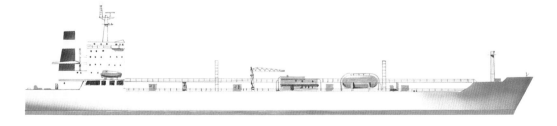

Small for a tanker, *Jakob Maersk* incorporated novel features, including bow and stern thrusters. Held in free-standing tanks in four holds, cargo is handled by eight multistage centrifugal pumps that can load or discharge two tanks at a time. The ship is now the LPG tanker *Maharshi Bhavatreya*.

SPECIFICATIONS

Type:	Danish tanker
Displacement:	42,523 tonnes (42,523 tons)
Dimensions:	185m x 27.4m x 12.5m (607ft x 90ft x 41ft)
Machinery:	Single screw, diesel engine
Top speed:	17.3 knots
Cargo:	Liquefied petroleum gas
Route:	Mediterranean Sea–Northern Europe
Complement:	23
Launched:	1991

TIMELINE　　　　1990　　　　　　　　　　　　1991

Front Driver

Front Driver is an OBO (oil/bulk ore) carrier. Nine holds with wing tanks can handle three types of oil cargo. The hull is made of 52 per cent high-tensile steel. The OBO concept was not widely followed, however. *Front Driver*, with 153,000 tonnes of coal, was the target of a Greenpeace demonstration in 2007.

SPECIFICATIONS	
Type:	Swedish cargo vessel
Displacement:	195,733 tonnes (192,651 tons)
Dimensions:	285m x 45m (935ft x 147ft 8in)
Machinery:	Single screw, diesel engine
Cargo:	iron ore, coal, limestone, salt, coke or grain in bulk
Routes:	International routes
Launched:	1991

Hakuryu Maru

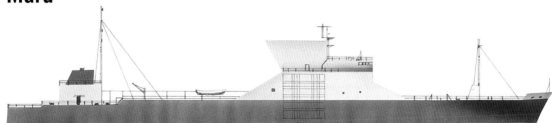

Hakuryu Maru was built specifically to handle and transport steel coils on pallet carriers. The hull has a double bottom with 1400 tonnes (1378 tons) of permanent iron and concrete ballast, to provide stable sea motion and, most importantly considering the loads involved, to limit heel to 3° while loading and unloading.

SPECIFICATIONS	
Type:	Japanese steel carrier
Displacement:	5278 tonnes (5195 tons)
Dimensions:	115m x 18m x 5m (377ft 4in x 59ft 16ft 5in)
Machinery:	Single screw, diesel engine
Top speed:	15 knots
Route:	Fukuyama-other Japanese ports
Launched:	1991

Futura

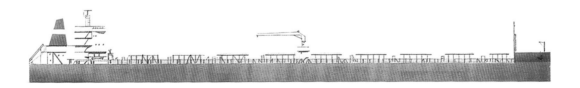

Futura was one of the first bulkers built to improved safety standards, including a double hull, gas ventilating of tanks and the ability to ventilate tanks selectively as needed via ballast pipes. Pumping facilities within the double hull eliminate the need for separate spaces to house this equipment in the cargo area.

SPECIFICATIONS	
Type:	Dutch bulk carrier
Displacement:	76,127 tonnes (74,928 tons)
Dimensions:	228.6m x 32.2m x 14.5m (750ft x 105ft 6in x 47ft 6in)
Machinery:	Single screw, diesel engine
Top speed:	14.8 knots
Cargo:	iron ore, coal grain and other bulk cargoes
Routes:	International routes
Launched:	April 1992

1992

Cargo Ships

Modern cargo ships follow set routes, though international shipping agencies exist to find ships for specific cargoes, or vice versa. Many ships carry little or no cargo-handling equipment, and their increasing size confines them to deep-water ports with dockside cranes and conveyors. Machinery and ship-handling are automated.

Savannah

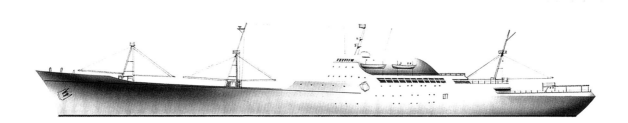

The first merchant ship powered by atomic propulsion, *Savannah* was meant to demonstrate the peaceful application of atomic energy. However, high operational costs made it commercially unviable and it was withdrawn from service in 1972. Its future as a museum ship was under debate in 2010.

SPECIFICATIONS

Type:	US cargo ship
Displacement:	14,112 tonnes (13,890 tons)
Dimensions:	195m (639ft 9in) x 23.77m (78ft)
Machinery:	Twin screws, nuclear reactor, turbines
Top speed:	20.5 knots
Cargo:	Mixed freight
Complement:	124
Launched:	July 1959

Helena

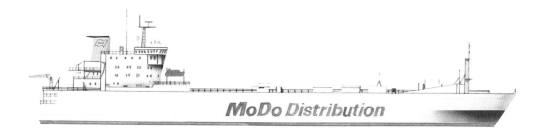

MoDo Distribution

Helena is a roll-on, roll-off freighter with a full-length double bottom, and a double skin in the lower cargo areas and engine room. Traffic when loading is two-way, and all four decks are interconnected by ramps. Cargo-handling is monitored by closed-circuit display units in the wheelhouse and the engine room control centre.

SPECIFICATIONS

Type:	Swedish cargo vessel
Displacement:	22,548 tonnes (22,193 tons)
Dimensions:	169m x 25.6m x 7m (554ft 6in x 84ft x 23ft)
Machinery:	Single screw, diesel engine
Top speed:	14.6 knots
Cargo:	Paper products, trailers, small cars and containers
Routes:	Swedish ports to Antwerp and other European ports
Launched:	1990

TIMELINE

1959   1990   1991

Hudson Rex

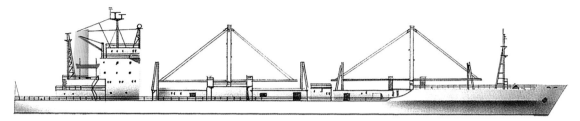

The design of *Hudson Rex* incorporates conventional derrick booms, operated by electro-hydraulic winches. Fans supply cold air to the refrigerated area, and there is a comprehensive insulation system. The refrigeration and temperature control system is installed in the engine room. The ship now sails as *Sun Maria*.

SPECIFICATIONS	
Type:	Panamanian cargo carrier
Displacement:	12,192 tonnes (12,000 tons)
Dimensions:	148.5m x 20.6m x 9.4m (487ft 3in x 67ft 7in x 31ft)
Machinery:	Single screw, diesel engine
Top speed:	19.2 knots
Cargo:	Refrigerated goods
Routes:	Netherlands–West Africa
Launched:	October 1991

Halla

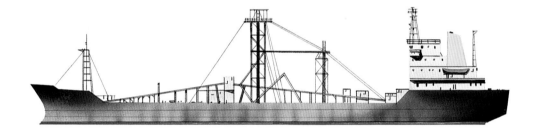

The ship is a self-loader and unloader, using compressed air power to move its cargo through large-diameter tubes. The lattice tower supports discharge booms, and the machinery room is between the two holds. 1000 tonnes (984 tons) of cement can be loaded, and 500 tonnes (492 tons) unloaded, in 1 hour.

SPECIFICATIONS	
Type:	Korean cement carrier
Displacement:	10,427 tonnes (10,427 tons)
Dimensions:	111.8m x 17.8m x 7m (367ft x 58ft 5in x 23ft)
Machinery:	Single screw, diesel engine
Top speed:	13 knots
Cargo:	bulk cement
Route:	South Korea–Japan
Complement:	27
Launched:	January 1991

Krasnograd

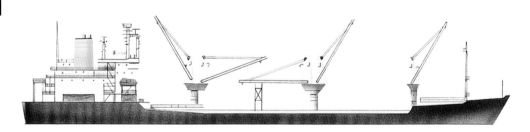

Krasnograd was one of the first foreign-built carriers to enter the Russian merchant fleet for years. The hull has two through-decks with four cargo holds. Containers of 6m (20ft) and 12m (40ft) offer a capacity of 728 TEU; 30 are refrigerated. The ship is now Greek-owned and Malta-registered as *Nordana Surveyor*.

SPECIFICATIONS	
Type:	Russian cargo ship
Displacement:	26,630 tonnes (26,630 tons)
Dimensions:	173.5m x 23m x 10m (569ft 3in x 75ft 6in x 32ft 10in)
Machinery:	Single screw, diesel engine
Cargo:	Containers
Routes:	International routes
Launched:	1992

1992

Jervis Bay

Jervis Bay moves between highly automated terminals, and unloading and reloading can be accomplished within 24 hours. The loading space is maximized by computer-assisted design, and on-board computers specify and record the location of every container carried on the ship. Carrying capacity is 4038 6m (20ft) units.

Jervis Bay

Despite its huge size, the *Jervis Bay* had a crew of just nine officers and 10 men. *Jervis Bay* and its sisters-ships were constructed to be employed mainly in the Europe–Far East service (accomplishing a round trip from Southampton to Yokohama and back in 63 days).

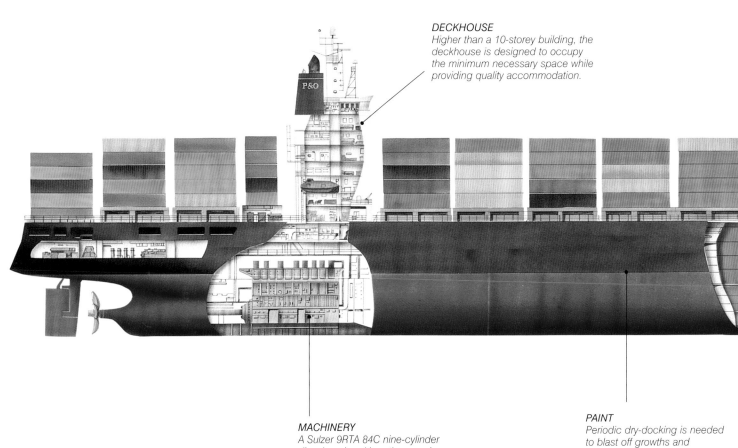

DECKHOUSE
Higher than a 10-storey building, the deckhouse is designed to occupy the minimum necessary space while providing quality accommodation.

MACHINERY
A Sulzer 9RTA 84C nine-cylinder diesel engine drives the massive propellor. Maximum power output is 34,412kW (47,000shp) at 100 rpm.

PAINT
Periodic dry-docking is needed to blast off growths and accretions and to repaint the hull with drag-reducing paint.

SPECIFICATIONS

Type:	British container ship
Displacement:	51,816 tonnes (51,000 tons)
Dimensions:	292.15m x 32.2m x 11.2m (985ft 6in x 105ft 6in x 36ft)
Machinery:	Single screw, diesel
Service speed:	23.5 knots
Cargo:	Containers
Routes:	Southampton–Japanese ports
Complement:	19
Launched:	1992

NAME
Passing into the ownership of A.P. Moller in 2006, and operated by Maersk Line, the ship was renamed MSC Almeria in 2008.

CONTAINERS
More than half the load is carried above deck level. 174 deck and 66 hold containers can be refrigerated.

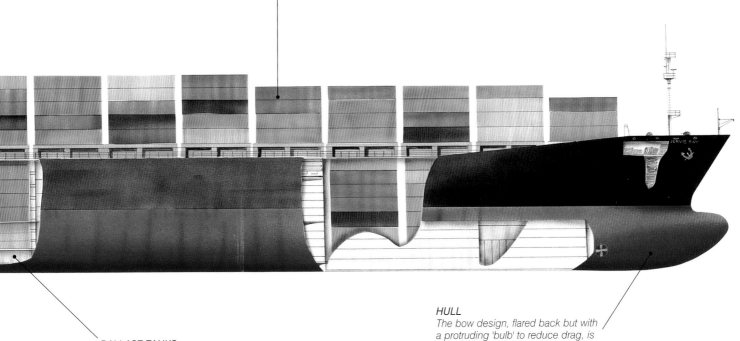

HULL
The bow design, flared back but with a protruding 'bulb' to reduce drag, is typical of the large modern motor ship.

BALLAST TANKS
Water pumped in and out of the ballast tanks is a vital necessity to maintain stability during loading and unloading.

Container Ships

The basic dimensions of all container ships are based on the standard international containers: 12m (40ft) and 6m (20ft) long. A ship designer's aim is to produce a hull that will hold the maximum number of loaded containers for the minimum usage of steel compatible with strength, safety and international marine regulations.

Ever Globe

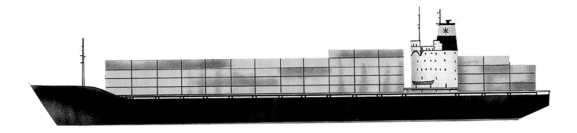

Ever Globe was first owned by the Evergreen Marine Corporation, based in Taiwan and registered in Panama. It had three holds for containers, plus a massive deck area where more could be stacked. Sold and renamed twice, it was *Scotland,* then *Hera.* In January 2009, it was sold for scrapping in China.

SPECIFICATIONS

Type:	Taiwanese container ship
Displacement:	43,978 tonnes (43,285 tons)
Dimensions:	231m x 32m (757ft 10in x 105ft)
Machinery:	Single screw, diesel engine
Routes:	Various
Launched:	1984

Hannover Express

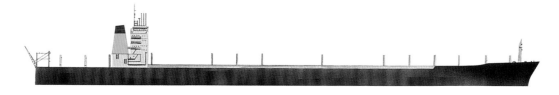

High-tensile steel construction allows 11 rows of containers to be stowed in the hull instead of 10. Additionally, rearranging the longitudinal beams allowed *Hannover Express* to carry heavy-lift cargoes, usually impossible with this type of vessel. The dimensions are 'Panamax', designed to fit the Panama Canal.

SPECIFICATIONS

Type:	German container ship
Displacement:	76,330 tonnes (75,128 tons)
Dimensions:	294m x 32.2m x 13.5m (964ft 6in x 105ft 10in x 44ft 4in)
Machinery:	Single screw, diesel engine
Top speed:	23.8 knots
Routes:	International routes
Complement:	21
Launched:	October 1990

TIMELINE

1984   1990   1991

Kota Wijaya

Registered in Singapore, *Kota Wijaya* carries a combined load of 6m (20ft) and 12m (40ft) containers in six holds, with a capacity of 1186 TEU, plus 200 refrigerated containers on the upper deck. The hull is double-skinned over the midships section. Heeling tanks with water ballast provide stability while loading.

SPECIFICATIONS	
Type:	Malayan container ship
Displacement:	22,695 tonnes (22,695 tons)
Dimensions:	184.5m x 27.6m x 9.5m (605ft 4in x 90ft 6in x 31ft 3in)
Machinery:	Single screw, diesel engines
Top speed:	19 knots
Routes:	Malaysia–Europe and Australia
Launched:	February 1991

Nedlloyd Europa

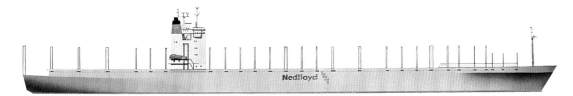

Nedlloyd Europa was the first ship to dispense with standard hatch covers, for a container guide-support system that extends from the holds up above the deck to secure deck-carried containers. Tests proved the safety of the design, which needs duplicate pumping and drainage systems. There are seven cargo holds.

SPECIFICATIONS	
Type:	Dutch container ship
Displacement:	48,768 tonnes (48,000 tons)
Dimensions:	266m x 32.2m x 13m (872ft 9in x 105ft 9in x 42ft 8in)
Machinery:	Single screw, diesel engines
Top speed:	23 knots
Routes:	International routes
Launched:	September 1991

Hyundai Admiral

With the world's most powerful diesel engine (developing over 67,000bhp), *Hyundai Admiral* is highly automated: control centre monitors enable a single watch-keeping system to be used. The seven holds accommodate over 4400 containers. There is an area for dangerous cargo. The ship has a double hull.

SPECIFICATIONS	
Type:	British container ship
Displacement:	62,131 tonnes (61,153 tons)
Dimensions:	275m x 37m x 13.6m (902ft 3in x 121ft 9in x 44ft 7in)
Machinery:	Single screw, diesel engines
Routes:	Britain–Far East
Launch date:	1992

1992

Anastasis

In 1978, the Italian liner *Victoria* was converted by the US Mercy Ships charity into a hospital ship, visiting countries where medical aid was in short supply. Fitted out to provide surgical and dental treatment, *Anastasis* visited 275 ports, on 66 field assignments, in 23 nations. After 29 years of service, it was retired in 2007.

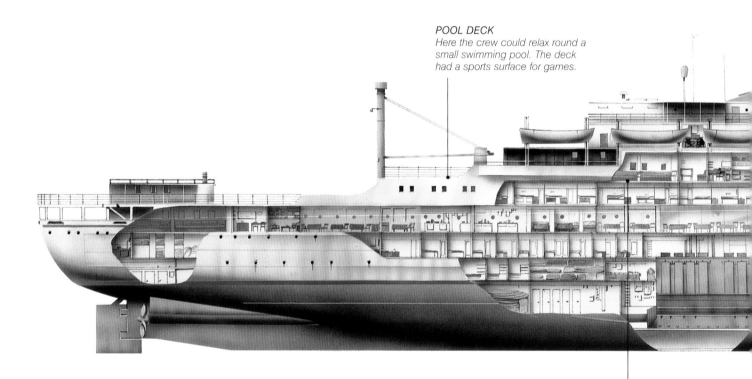

POOL DECK
Here the crew could relax round a small swimming pool. The deck had a sports surface for games.

CT SCANNER
A CT (computerised tomography) scanner was added to the ship's range of diagnostic and treatment facilities in 2002.

Anastasis

As a medical vessel *Anastasis* has delivered aid to some of the world's poorest countries. In the previous role as a passenger ship it sailed to Pakistan, India and the Far East.

SPECIFICATIONS

Type:	Italian liner, later US private hospital ship
Displacement:	11,882 tonnes (11,695 tons)
Dimensions:	159m x 20.7m (521ft 6in x 67ft 10in)
Machinery:	Twin screws, diesels
Top speed:	19.5 knots
Complement:	420
Launched:	1953

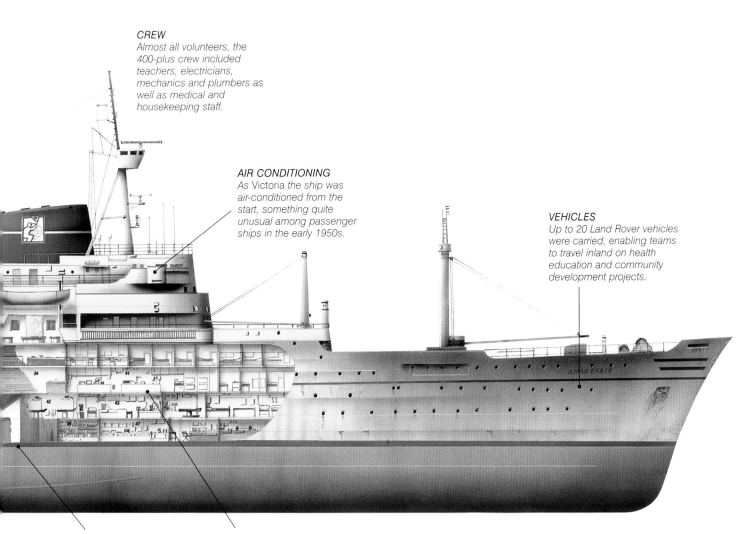

CREW
Almost all volunteers, the 400-plus crew included teachers, electricians, mechanics and plumbers as well as medical and housekeeping staff.

AIR CONDITIONING
As Victoria *the ship was air-conditioned from the start, something quite unusual among passenger ships in the early 1950s.*

VEHICLES
Up to 20 Land Rover vehicles were carried, enabling teams to travel inland on health education and community development projects.

MACHINERY
Two Fiat 7510 diesel engines each developing 5,995kW (8,040bhp). Much of the ship's secondary machinery had to be updated or replaced.

MEDICAL SUITES
Three operating theatres, a 40-bed ward for recuperation, and a two-bed intensive care unit were set up.

Cruise Ships

Cruising as a holiday had been possible for more than a century, but took place on a limited scale. The availability of big ships, with adroit marketing and refits that included more entertainment and leisure facilities, combined to turn cruising into an all-year round marine industry. Soon custom-built cruise ships would be launched.

Galileo Galilei

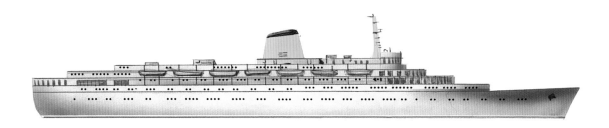

Built for the Italy–Australia route, but made redundant by the Jumbo Jet, *Galileo Galilei* was laid up from 1977 to 1979. From 1983, it was the cruise ship *Galileo,* rebuilt in 1983 as *Meridian,* with gross tonnage increased to 30,500 tonnes. Sold in 1996 and renamed *Sun Vista,* it sank off Malaysia in May 1999 after a fire.

SPECIFICATIONS

Type:	Italian liner, later cruise ship
Displacement:	28,353.5 tonnes (27,907 tons)
Dimensions:	213.65m x 28.6m x 8.65m (700ft 11in x 93ft 10in x 28ft 4in)
Machinery:	Twin screws, geared turbines
Top speed:	24 knots
Launched:	1963

Radisson Diamond

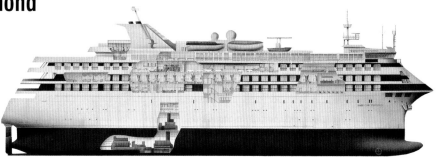

The big passenger ship found a new role as a cruise liner, but design changed radically. This ship is small compared to later examples, its catamaran design making it broad in relation to its length. Each hull has an engine, and boats can be launched from between them. Now *Asia Star,* it operates from Hong Kong.

SPECIFICATIONS

Type:	US cruise ship
Displacement:	18,684 tonnes (18,400 tons)
Dimensions:	131m x 32m (423ft x 105ft)
Machinery:	Twin screws plus bow thrusters, diesels
Top speed:	12.5 knots
Launched:	1991

TIMELINE

1963  1991

Society Adventurer

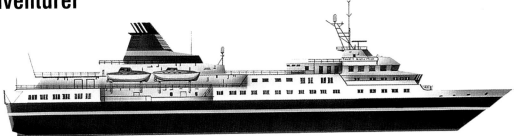

This expedition ship was renamed *Hanseatic* by new German owners in 1993. It can operate for up to eight weeks without taking on fuel or provisions. Range is 16,150km (8500 miles). For cruises to special-interest areas such as Antarctica, it has an observation lounge above the bridge, and carries 188 passengers.

SPECIFICATIONS	
Type:	Bahamian expedition ship
Displacement:	8,512 tonnes (8,378 tons)
Dimensions:	122.7m x 18m x 4.7m (402ft 8in x 59ft x 15ft 8in)
Machinery:	Twin screws, diesel engines
Launched:	January 1991

Majesty of the Seas

Built for Royal Caribbean Cruise Lines, *Majesty of the Seas* has promenade decks that are largely enclosed. It carries 2350 passengers on week-long cruises in the Caribbean Sea. Engines are low down and aft, maximizing passenger space and minimizing vibration and noise. It was partly refurbished in 2007.

SPECIFICATIONS	
Type:	Norwegian cruise ship
Displacement:	75,124 tonnes (73,941 tons)
Dimensions:	266.4m x 32.3m x 7.6m (874ft x 106ft x 25ft)
Machinery:	Single screw, diesels
Top speed:	21 knots
Launched:	1992

Europa

Rated 'best cruise ship' for nine successive years, *Europa* accommodates only 410 passengers. Novel aspects of the design include a propeller drive from externally mounted 'pods' that can be angled to increase manoeuvrability, and the propellers themselves are pullers rather than pushers. It was refitted in 2007.

SPECIFICATIONS	
Type:	German cruise ship
Displacement:	28,854.4 tonnes (28,400 tons)
Dimensions:	198.6m x 78ft x 39ft (644ft 6in x 78ft x 39ft)
Machinery:	Twin screws, diesel-electric
Top speed:	21 knots
Launched:	1999

1992    1999

Canberra

Canberra served with P&O on the Pacific route in May 1961. Lavishly fitted, it could carry 2186 passengers. Requisitioned in 1982 as a troop ship for the Falklands campaign, it had several narrow escapes. It returned to the UK in July 1982 and, after a refit, returned to service. *Canberra* was scrapped in Pakistan in 1997–8.

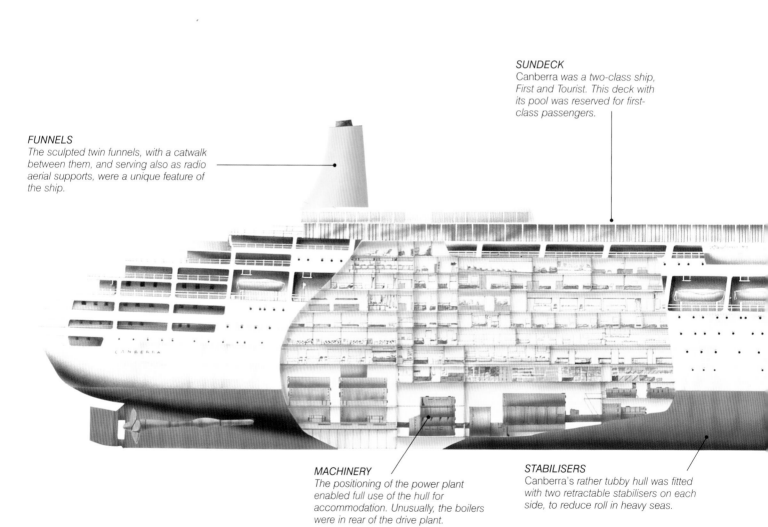

SUNDECK
Canberra *was a two-class ship, First and Tourist. This deck with its pool was reserved for first-class passengers.*

FUNNELS
The sculpted twin funnels, with a catwalk between them, and serving also as radio aerial supports, were a unique feature of the ship.

MACHINERY
The positioning of the power plant enabled full use of the hull for accommodation. Unusually, the boilers were in rear of the drive plant.

STABILISERS
Canberra's *rather tubby hull was fitted with two retractable stabilisers on each side, to reduce roll in heavy seas.*

Canberra

Canberra was built in Belfast, Northern Ireland in 1960 and entered service the following year. It was originally designed to serve as a passenger liner between Great Britian and Australia, but it was converted into a luxury cruise ship in 1974.

SPECIFICATIONS	
Type:	British liner
Displacement:	45,524 tonnes (44,807 tons)
Dimensions:	249m x 31m (817ft x 101ft 8in)
Machinery:	Twin screws, steam turbines, electric drive
Route:	Britain–Australia
Complement:	938
Launched:	March 1960

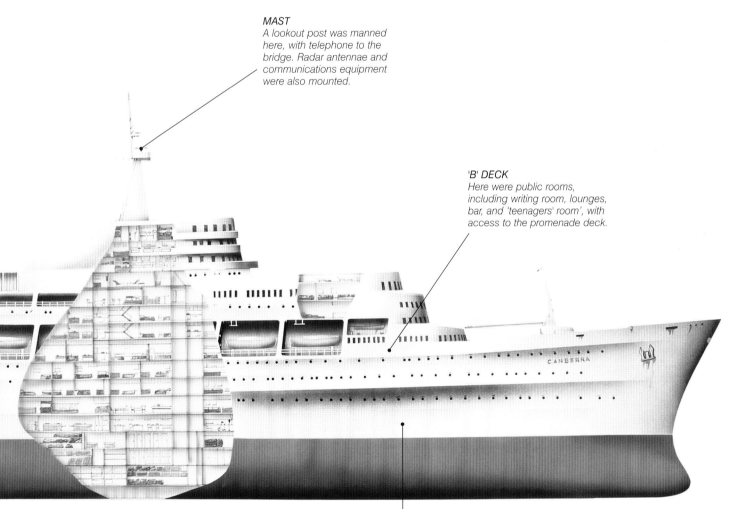

MAST
A lookout post was manned here, with telephone to the bridge. Radar antennae and communications equipment were also mounted.

'B' DECK
Here were public rooms, including writing room, lounges, bar, and 'teenagers' room', with access to the promenade deck.

CREW ACCOMMODATION
Crew cabins, some tourist class cabins, baggage rooms, store rooms and workshops were situated on 'G' deck at waterline level.

Queen Elizabeth II

Queen Elizabeth II was the last of the transatlantic liners. Construction was beset by problems, and mechanical troubles delayed its maiden voyage. It was requisitioned as a troopship in the Falklands War of 1982. Re-engined, it later served successfully as a cruise ship. Withdrawn in 2008, it was bought to be a floating hotel at Dubai.

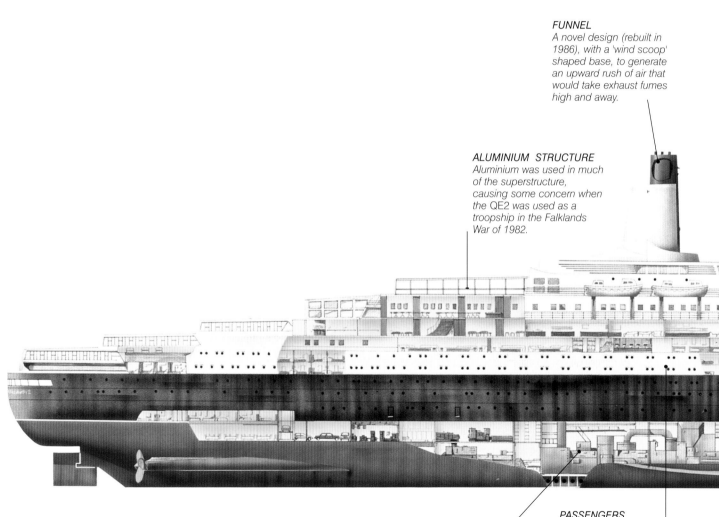

FUNNEL
A novel design (rebuilt in 1986), with a 'wind scoop' shaped base, to generate an upward rush of air that would take exhaust fumes high and away.

ALUMINIUM STRUCTURE
Aluminium was used in much of the superstructure, causing some concern when the QE2 was used as a troopship in the Falklands War of 1982.

MACHINERY
The original steam turbine machinery was wholly replaced in 1986-87 by nine MAN 9-cylinder diesels with electric drive.

PASSENGERS
In a two-class arrangement, First Class held 548 passengers and Tourist held 1690. Officers and crew numbered 795.

Queen Elizabeth II

In May 1982 the *QE2* took part in the Falklands War, carrying 3000 troops and 650 volunteer crew to the South Atlantic. It was refitted in Southampton in preparation for war service, this included the installation of three helicopter pads, the transformation of public lounges into dormitories and the covering of carpets with 2000 sheets of hardboard.

SPECIFICATIONS

Type:	British passenger liner
Displacement:	66,432 tonnes (65,836 tons)
Dimensions:	293.5m x 32.1m x 9.75m (963ft x 105ft x 32ft)
Machinery:	Twin screws, geared turbines (later diesel-electric)
Top speed:	29 knots
Routes:	North Atlantic; cruising
Launched:	1967

BALCONIES
New penthouse accommodation was built behind the bridge in 1977, offering private balconies to residents.

PROFILE
The aim was to retain as much of the classic 'Cunarder' look as possible, though with a single mast and funnel.

A PROUD RECORD
On retirement the QE2 had sailed almost 9.656 million km (6 million miles), carried 2.5 million passengers and crossed the Atlantic 806 times.

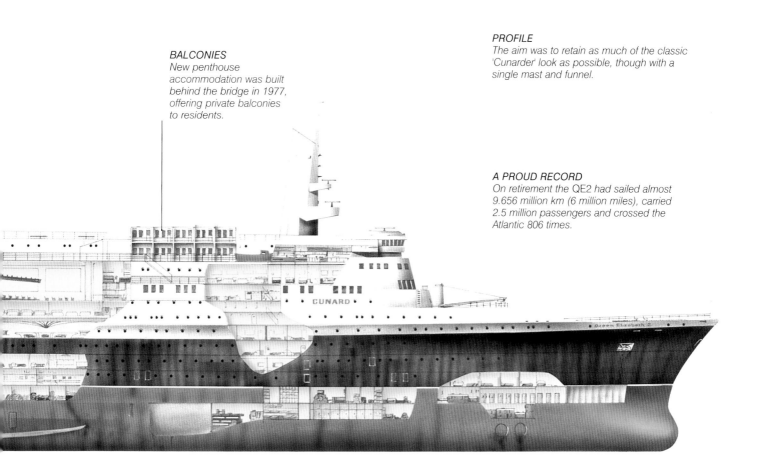

Seagoing Ferries

Trends in travel and transport have encouraged the concept of the big, long-range car and truck ferry with comfortable accommodation and many amusements and attractions for passengers. The most notable examples ply the Baltic Sea and Japan–Asia routes and the tendency is for such vessels to get ever bigger.

Ishikari

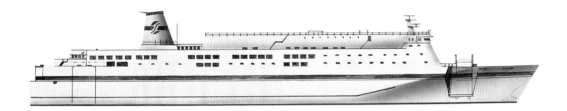

Ishikari was one of Japan's first luxury, high-speed ferries, modelled on Baltic Sea ferries, and is able to carry 850 passengers, 151 cars and 165 trucks. The hull is designed for speed and economy, and it burns 76 tonnes (75 tons) of oil a day in regular service. The upper part houses nine separate decks.

SPECIFICATIONS

Type:	Japanese vehicle ferry
Displacement:	7050 tonnes (6938 tons)
Dimensions:	192.5m x 27m x 6.9m (631ft 6in x 88ft 7in x 22ft 8in)
Machinery:	Twin screws, diesel engines
Top speed:	21.5 knots
Route:	Between Japanese islands
Launched:	November 1990

Ferry Lavender

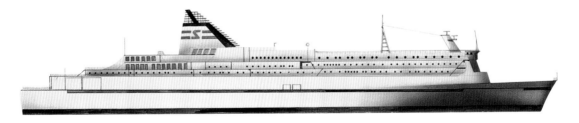

A roll-on, roll-off passenger ferry, ordered by the Shin Nihonkai Ferry Company, *Ferry Lavender* was one of the biggest car ferries of its day. Able to carry 796 passengers, its two vehicle decks have access through bow and stern ramps. Greek-owned from 2004 as *Ionian King*, it now works between Greece and Italy.

SPECIFICATIONS

Type:	Japanese vehicle ferry
Displacement:	20,222 tonnes (19,904 tons)
Dimensions:	193m x 29.4m x 6.7m (632ft x 96ft 5in x 22ft 2in)
Machinery:	Twin screws, diesel engines
Top speed:	21.8 knots
Route:	Patras–Brindisi
Launched:	March 1991

TIMELINE

1990  1991

Tycho Brahe

The largest double-ended train ferry, *Tycho Brahe* works between Helsingør, Denmark, and Helsingborg, Sweden. It was designed to reach maximum speed after 1500m (1640yd), and to decelerate rapidly when 800m (875yd) from shore. It carries 260 lorries, 240 cars and nine railway coaches, and 1250 passengers.

SPECIFICATIONS	
Type:	Danish train ferry
Displacement:	10,871 tonnes (10,700 tons)
Dimensions:	111m x 28.2m x 5.7m (364ft 2in x 92ft 6in x 18ft 8in)
Machinery:	Quadruple thrusters, diesel engines
Top speed:	13.5 knots
Route:	Helsingør–Helsingborg
Launched:	1991

Frans Suell

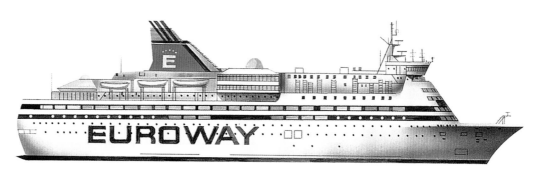

From the keel of this Baltic Sea 'super-ferry' to the top of the wheelhouse, there are 12 decks, two of which are used to hold road vehicles. Accommodation is provided for 2300 passengers, with some cabins having balconies. Ownership passed to Silja as *Silja Scandinavia*, then to Viking Line as *Gabriella*.

SPECIFICATIONS	
Type:	Swedish ferry
Displacement:	35,850 tonnes (35,285 tons)
Dimensions:	169.4m x 27.6m x 6.25m (556ft x 90ft 6in x 20ft 6in)
Machinery:	Twin screws, diesel engines
Route:	Stockholm–Helsinki
Launched:	January 1991

Condor Express

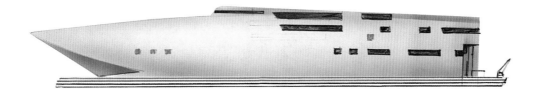

Built at the Condor yard in Hobart, Tasmania, this catamaran-hull ferry takes up to 200 cars and 776 passengers. Its four Ruston 20-cylinder diesel engines propel it at speeds in excess of 40 knots. With two sister craft, it provides a fast service between England's south coast, the Channel Islands and Brittany.

SPECIFICATIONS	
Type:	British car ferry
Displacement:	386 tonnes (380 tons)
Dimensions:	86.25m x 26m x 3.5m (282ft 10in x 85ft 3in x 11ft 6in)
Machinery:	Four water jets, diesels
Top speed:	40 knots
Route:	Poole–St Malo
Launched:	1997

1997

Multi-Product & Oil Tankers

Some disastrous groundings and oil spillages prompted a tighter international regime for construction and operation of VLCC (very large crude carrier) ships. Maximum permitted Panama Canal (Panamax) dimensions of 294.13m x 32.31m x 12.04m (965ft x 106ft x 39ft 6in) will be increased by 2014, enabling increases in ship sizes.

British Skill

British Skill was one of several ships built to replace ageing supertankers of previous decades. Like all big modern ships, controls are largely automated. It was an early user of Doppler radar for assisting with steering at low speed. In 2000, it was subject to a pirate attack in the South China Sea.

SPECIFICATIONS

Type:	British tanker
Displacement:	67,090 tonnes (66,034 tons) gross, 129,822 tonnes (127,778 tons) deadweight
Dimensions:	261m x 40m (856ft 5in x 131ft 5in)
Machinery:	Twin screws, diesel engines
Cargo:	Crude oil
Complement:	30–40
Launched:	1980

Mayon Spirit

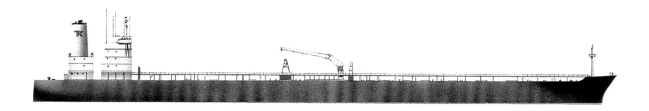

Mayon Spirit has a double hull with a 2m (6ft 6in) space in the double bottom and more space in the wing tanks. There is a central cargo tank in a midship position and small side tanks. Carrying capacity is 120,043 cubic metres (4,239,285 cubic feet). Cargo is handled by three pumps monitored from the control room.

SPECIFICATIONS

Type:	Liberian tanker
Displacement:	100,000 tonnes (98,507 tons)
Dimensions:	244.8m x 41.2m x 14.4m (830ft 2in x 135ft 2in x 47ft 3in)
Machinery:	Single screw, diesel engine
Cargo:	crude oil
Complement:	38
Launched:	December 1981

TIMELINE

1980   1981   1990

Helice

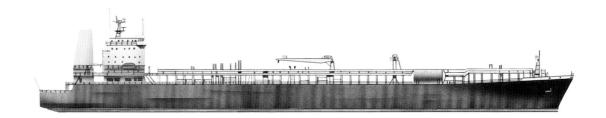

The four cargo holds have free-standing prismatic tanks built from carbo-manganese steel. Two control, or purge, tanks are situated on deck. The air can be changed up to eight times per hour in the largest hold if necessary. Now owned by Varun Line of India, it is renamed *Maharshi Vamadeva* and carries LPG.

SPECIFICATIONS	
Type:	Norwegian tank vessel
Displacement:	50,292 tonnes (49,500 tons)
Dimensions:	205m x 32.2m x 13m (672ft 7in x 105ft 8in x 42ft 8in)
Machinery:	Single screw, diesel engine
Top speed:	16 knots
Cargo:	liquefied petroleum gas
Launched:	September 1990

Landsort

Landsort was the first double-hulled oil tanker to comply with new regulations for this type. In 1997 Greek owners renamed it *Crudegulf*; in 2003 it became *Genmar Gulf*. Nine full-width tanks give capacity of 172,850 cubic metres (6,105,150 cubic ft). The space between the hulls is divided into 10 water ballast tanks.

SPECIFICATIONS	
Type:	Swedish tanker (VLCC)
Displacement:	165,646 tonnes (163,038 tons)
Dimensions:	274m x 48m x 17m (899ft x 157ft 6in x 55ft 9in)
Machinery:	Single screw, diesel engine
Cargo:	crude oil and oil products of heavier specific gravity
Launched:	June 1991

Jo Alder

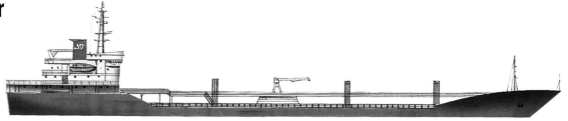

This specialized tanker for food products, non-contaminating chemicals and general petroleum products was renamed *Valdarno*, then *Monte Chiaro*. The hull is double-skinned and double-bottomed, with toughened longitudinal and transverse bulkheads. The engine room is designed for unmanned operation.

SPECIFICATIONS	
Type:	Italian tanker
Displacement:	12,801 tonnes (12,600 tons)
Dimensions:	139m x 21.2m x 8m (456ft x 69ft 9in x 26ft 5in)
Machinery:	Single screw, diesel engine
Top speed:	14.5 knots
Cargo:	Liquids
Launched:	1991

1991

Specialized Vessels

The spread of undersea oil drilling from shallow to deep water has extended the range of specialized surface ships needed to provide support vessels and to transport large structures. The spread of undersea fibre-optic and high-tension electricity cables has prompted a new generation of high-tech cable-laying ships.

Endurance

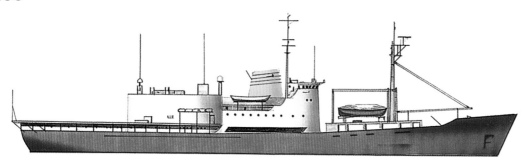

Anita Dan was built for the Lauritzen Line and renamed in 1967, when purchased by Britain for conversion to an ice patrol vessel. It entered service as support ship to the British South Atlantic Survey in 1968, and had a major refit in 1978. Weakened after hitting an iceberg in 1989, it was withdrawn in 1991.

SPECIFICATIONS	
Type:	British ice patrol ship
Displacement:	3657 tonnes (3600 tons)
Dimensions:	91.5m x 14m x 5.5m (300ft x 46ft x 18ft)
Machinery:	Single screw, diesel engine
Top speed:	14.5 knots
Main armament:	Two 20mm (0.79in) guns
Launched:	May 1956

Batcombe

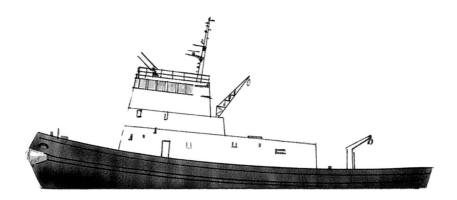

A small tug-boat, *Batcombe* was equipped with foam and water tanks and high-pressure hose equipment to deal with fires on ships or in quayside installations. The hoses are mounted above the control room and the deckhouse has a rounded structure to give maximum field of access.

SPECIFICATIONS	
Type:	British fire-fighting tug
Dimensions:	18m x 5.4m (60ft x 18ft)
Machinery:	Single screw, diesel engines
Launched:	1970

TIMELINE

1956  1970 1982

AP.1-88

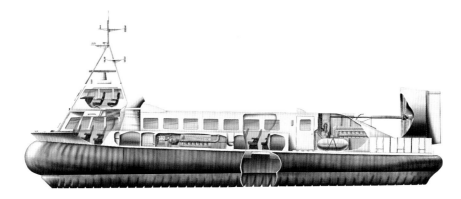

Advances in diesel engine design let hovercraft dispense with costly gas turbines, becoming commercially viable. This was the first commercial class, fitted as a passenger or cargo craft, able to carry 101 passengers or 12 tonnes of freight. A military version was produced, capable of mounting light cannon or missiles.

SPECIFICATIONS	
Type:	British air cushion vehicle
Maximum operating weight:	40.6 tonnes (40 tons)
Dimensions:	24.5m x 11m (80ft 4in x 36ft)
Machinery:	Four diesel engines driving air cushion fan
Top speed:	40 knots
Launched:	1982

KDD Ocean Link

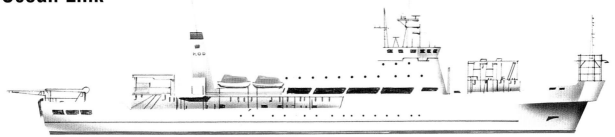

KDD Ocean Link can operate in the severe conditions of the North Pacific. Three large holds store the cables. High-speed Optical Fibre Cable is laid over the stern, while cable trenching is accomplished by a towed burial plough, fitted with TV camera and scanning sonar, and is monitored from the main control room.

SPECIFICATIONS	
Type:	Japanese cable layer
Displacement:	9662 tonnes (9510 tons)
Dimensions:	133m x 19.6m x 7.4m (437ft x 64ft 4in x 24ft 3in))
Machinery:	Twin screws, diesel engines
Launched:	August 1991

Sea Spider

Renamed *Team Oman*, it was built to lay the submarine cable between Sweden and Poland. It carries 5000 tonnes (4921 tons) of high-voltage cable on a 24m (78ft) diameter carousel, and 1600 tonnes (1574 tons) in a cable basket. Three cable engines with extending arms are mounted. A stern frame supports the trenching gear.

SPECIFICATIONS	
Type:	Dutch cable layer
Displacement:	4072 tonnes (4008 tons)
Dimensions:	86.1m x 24m x 4.5m (285ft x 78ft x 15ft)
Machinery:	Controllable pitch screw, diesel electric
Top speed:	9.5 knots
Launched:	1999

1991   1999

James Clark Ross

Designed to carry out oceanographic research in the Antarctic region, with a strengthened bow and hull, *James Clark Ross* can pass through broken ice up to 1.5m (5ft) thick, or fragmented ice over 3m (10ft) thick. Capable of remaining at sea for up to 10 months at a time, the ship has fully equipped laboratory facilities.

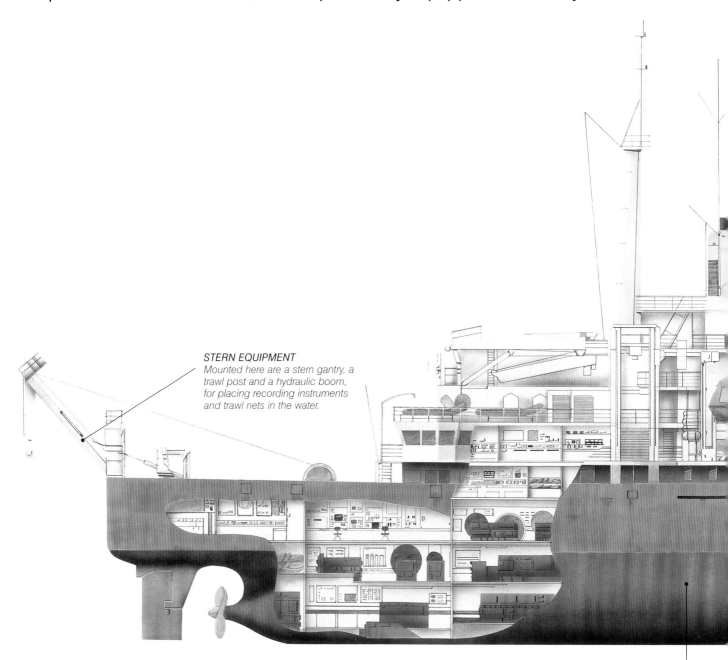

STERN EQUIPMENT
Mounted here are a stern gantry, a trawl post and a hydraulic boom, for placing recording instruments and trawl nets in the water.

HULL
A compressed air system enables the hull to roll from side to side, further breaking ice around the vessel.

James Clark Ross

This vessel is named after Admiral Sir James Clark Ross and is equipped for geophysical studies. It is designed with an extremely low noise signature to allow sensitive underwater acoustic equipment to operate effectively.

SPECIFICATIONS

Type:	British research vessel
Displacement:	7439 tonnes (7322 tons)
Dimensions:	99m x 10.8m x 6.5m (325ft x 35ft 5in x 21ft 4in)
Machinery:	Single screw, diesel engines
Launched:	December 1990

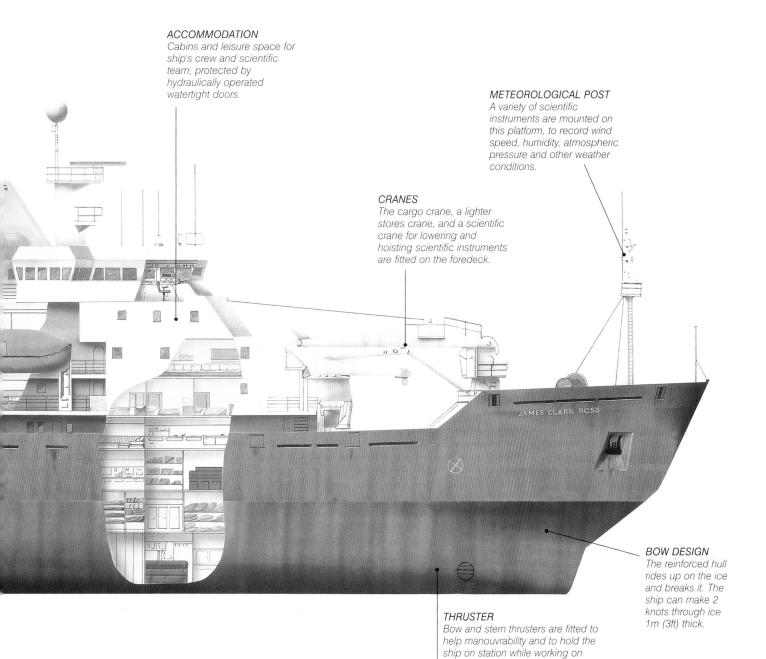

ACCOMMODATION
Cabins and leisure space for ship's crew and scientific team, protected by hydraulically operated watertight doors.

METEOROLOGICAL POST
A variety of scientific instruments are mounted on this platform, to record wind speed, humidity, atmospheric pressure and other weather conditions.

CRANES
The cargo crane, a lighter stores crane, and a scientific crane for lowering and hoisting scientific instruments are fitted on the foredeck.

BOW DESIGN
The reinforced hull rides up on the ice and breaks it. The ship can make 2 knots through ice 1m (3ft) thick.

THRUSTER
Bow and stern thrusters are fitted to help manouvrability and to hold the ship on station while working on geophysical projects.

Twentieth-Century Unarmed Submersible Craft

A range of scientific, technical and military needs drove research into submarines that could go very deep or use long-term power sources. Sea-bed exploration and mapping is increasingly important. Remote-controlled craft are often used.

Trieste

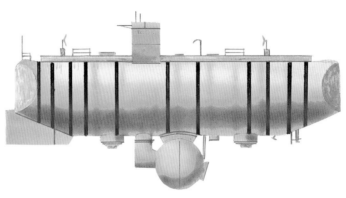

Designed by Auguste Piccard, *Trieste* was a large tank with a small crew sphere below. The tank held gasoline, lighter than water, enabling it to rise to the surface when water ballast was pumped out. In 1958, Piccard sold it to the US Navy, and in January 1960 it reached a record depth of 10,912m (35,800ft).

SPECIFICATIONS

Type:	French bathyscaphe
Displacement:	50.8 tonnes (50 tons)
Dimensions:	18.1m x 3.5m (59ft 6in x 11ft 6in)
Machinery:	Twin screws, electric motor
Top speed:	1 knot
Launched:	1953

Aluminaut

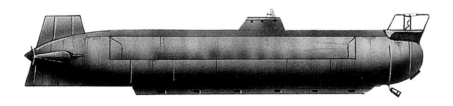

Constructed of aluminium 170mm (6.5in) thick, *Aluminaut* could attain depths of 4475m (14,682 ft). A side-scan sonar built up a map of the terrain on either side. *Aluminaut* was used in searching out the US H-bomb lost off Spain in 1966. Decommissioned in 1970, it is on exhibition at Richmond, Virginia.

SPECIFICATIONS

Type:	US deep sea exploration vessel
Displacement:	Unknown
Weight:	81 tonnes (80 tons)
Length:	16m (51ft)
Top speed:	3 knots
Launched:	1965

TIMELINE

1953 　　1965 

Deepstar 4000

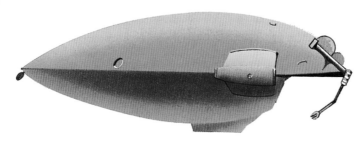

Deepstar 4000 was built between 1962 and 1964 by the Westinghouse Electric Corporation and the Jacques Cousteau group OFRS, as a scientific research and exploration submersible. The steel hull has 11 openings, and the two drive motors are attached externally. The craft carried a range of scientific equipment.

SPECIFICATIONS	
Type:	US submarine research craft
Displacement:	Unknown
Dimensions:	5.4m x 3.5m x 2m (17ft 9in x 11ft 6in x 6ft 6in)
Machinery:	Two fixed, reversible five hp AC motors
Top speed:	3 knots
Launched:	1965

Deep Quest

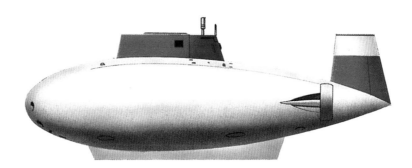

Deep Quest was built with fairing around a double sphere, one for the crew, the other for the propulsion unit. A deep search and recovery submarine, it could reach a depth of 2438 metres (8000ft). Vessels of this type remain vital when examining the seabed for sunken objects, pipelines and mineral deposits.

SPECIFICATIONS	
Type:	US submarine recovery craft
Displacement:	5 tonnes (5 tons)
Dimensions:	12m (39ft 4in) long
Machinery:	Twin reversible thrust electric motors
Service speed:	4.5 knots
Launched:	June 1967

India

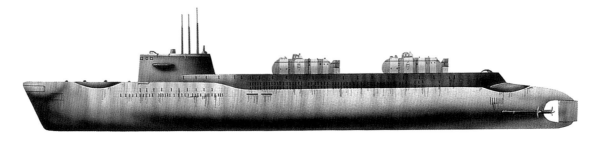

Project 940 Lenok was designed for salvage and rescue operations, under sea or ice. Two DSRV (deep submergence rescue vessels) were carried in semi-recessed deck wells aft, and could link with the submerged parent boat. Two boats formed the class, both withdrawn in 1990 and scrapped in 1995.

SPECIFICATIONS	
Type:	Soviet rescue submarine
Displacement:	3251 tonnes (3200 tons) [surface], 4064 tonnes (4000 tons) [submerged]
Dimensions:	106m x 10m (347ft 9in x 32ft 10in)
Machinery:	Twin screws, diesel engines [surface], electric motors [submerged]
Top speed:	15 knots [surface], 11 knots [submerged]
Complement:	94
Launched:	1979

1967

1979

TWENTY-FIRST CENTURY SHIPS

Much attention is now given to reducing ships' environmental impact, both on the water and on the atmosphere. This applies both to construction and to daily operation.

Rising oil prices also push operators towards more efficient use of fuel. Overall ship dimensions are set to increase after widening of the Panama Canal and the deepening of the Suez Canal. The enormous cost of modern warships has led to a reduction in size of most navies, though a 'destroyer' of the 2010s has more destructive capacity than a battleship of the 1940s.

Left: The *Type 45* is the largest and most powerful destroyer ever built for Britain's Royal Navy, with a destructive power undreamt of when the first 'torpedo-boat destroyers' were built.

Twenty-First Century Warships: 1

The US Navy has focused its attention on two key types of surface ship. First is the large aircraft carrier, capable of providing a powerful platform for air strikes anywhere in the world's seas. Second is the amphibious assault ship, equipped to transport and land combat troops and their mechanized and armoured support vehicles.

Charles de Gaulle

The only non-US nuclear-powered carrier, flying Super Etendard, Rafale, and E-2C Hawkeye aircraft. Propeller problems delayed commissioning until 2001 and reduced its operating speed. It saw active service in Operation Enduring Freedom in 2001-2, flying missions against al-Qaeda targets. It had a 15-month refit in 2007-8 including new propellers, but required further repairs in 2009.

SPECIFICATIONS

Type:	French aircraft carrier
Displacement:	42,672 tonnes (42,000 tons)
Dimensions:	261.5m x 64.36m x 9.43m (858ft x 211ft 2in x 30ft 10in)
Machinery:	Twin screws, two pressurised water nuclear reactors, four diesel-electric motors
Top speed:	27 knots
Main armament:	Four SYLVER launchers, MBDA Aster SAM, Mistral short-range missiles
Aircraft:	40
Complement:	1350 plus 600 air wing
Launched:	1994

Ronald Reagan

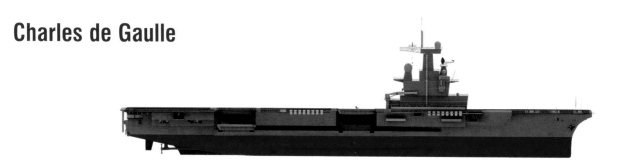

This ninth and penultimate ship in the *Nimitz* class has many new design features. Its island is placed further aft to give more open flight deck. Air traffic control and instrument landing guidance are included in a comprehensive range of systems. In 2008 *Reagan* launched over 1,140 sorties into Afghanistan. Its home port is San Diego, California.

SPECIFICATIONS

Type:	US aircraft carrier
Displacement:	103,000 tonnes (101,000 tons)
Dimensions:	332.8m x 76.8m x 11.3m (1092ft x 252ft x 37ft)
Machinery:	Quadruple screws, two nuclear reactors, four turbines
Top speed:	30+ knots
Main armament:	Two Mk 29 ESSM launchers, two RIM-116 Rolling Airframe Missile launchers
Armour:	Classified
Aircraft:	90 aircraft
Complement:	3200 plus 2480 air wing
Launched:	March 2001

TIMELINE

1994  2001  2004

Mistral

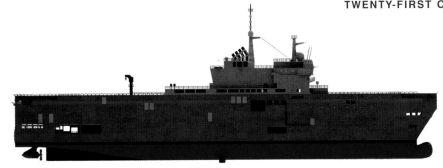

Built partly in St Nazaire and partly in Brest, where it was completed, *Mistral* was commissioned in February 2006, and deployed off Lebanon later that year. For short-range missions, troop capacity can double to 900. Two landing barges and 70 vehicles are also carried. In 2010 it was announced that Russia was to buy a *Mistral*-type ship.

SPECIFICATIONS	
Type:	French amphibious assault ship
Displacement:	21,300 tonnes (20,959 tons) full load
Dimensions:	199m x 32m x 6.33m (652ft 9in x 105ft x 20ft 8in)
Machinery:	Twin screws, diesel-electric motors
Top speed:	18.8 knots
Main armament:	Two Simbad systems
Aircraft:	16 heavy or 35 light helicopters
Complement:	310 plus 450 troops
Launched:	Ooctober 2004

New York

Steel salvaged from the World Trade Centre was incorporated in the construction of this fifth ship in the *San Antonio* class, though it is actually named for NY state. It carries two air-cushion attack craft (LCAC) or one LCU (Landing Craft Utility). Delivered in August 2009, it has been troubled by main bearing failures in its engines.

SPECIFICATIONS	
Type:	US amphibious transport dock ship
Displacement:	25,298.4 tonnes (24,900 tons) full load
Dimensions:	208.5m x 31.9m x 7m (684ft x 105ft x 23ft)
Machinery:	Twin screws, turbo diesels
Top speed:	22 knots
Main armament:	Two Bushmaster II 30mm (0.18in) Close In guns, two RAM missile launchers
Aircraft:	Two CH-53E Super Stallion, two MV-22B Osprey tiltrotor aircraft, four CH-46 Sea Knight, four AH-1 Sea Cobra or UH-1 Iroquois helicopters.
Complement:	360 plus 700 marines
Launched:	December 2007

Zumwalt

New and still developing technologies are incorporated in this $1.1 billion-plus ship. Its hull, recalling the old ironclads, is claimed to have the radar print of a fishing boat. Automated systems in every department reduce crew numbers dramatically. Land attack as much as sea-based targets is envisaged as a potential mission. Three are currently under construction.

SPECIFICATIONS	
Type:	US multimission destroyer
Displacement:	14,797 tonnes (14,564 tons)
Dimensions:	182.9m x 24.6m x 8.4m (600ft x 80ft 8in x 27ft 7in)
Machinery:	Gas turbines, emergency diesels, powering advanced induction motors (AIM)
Top speed:	30.3 knots
Main armament:	20 Mk 57 VLS modules, Evolved Sea Sparrow and Tomahawk missiles, two 155mm (2.24in) advanced gun systems, two 57mm (6.1in) guns (CIWS)
Sensors:	AN/SPY multi-function radar, volume search radar, dual band sonar
Aircraft:	1 helicopter
Complement:	140
Launch date:	2015

2007  2015

Twenty-First Century Warships - 2

It takes great resources and a substantial industrial and naval infrastructure to sustain the design, building and operation of modern warships. Increasingly, navies and constructors of allied powers are joining forces to plan and fund the development of new warships: a trend particularly noticeable among the countries of the European Union.

Sachsen

The *Sachsen* class air-defence frigates are an enhanced version of the *Brandenburg* class and share features with the Spanish F100 class. Extensive countermeasures equipment include chaff and flare launchers, and detection systems include long-range air and surface surveillance and target indication radar. An STN Atlas Elektronik DSQS-24B bow sonar is fitted.

SPECIFICATIONS

Type:	German frigate
Displacement:	5690 tonnes (5599 tons)
Dimensions:	143m x 17.4m x 5m (469ft x 57ft x 16ft 4in)
Machinery:	Twin controllable pitch screws, gas turbines and diesels (CODAG)
Top speed:	29 knots
Main armament:	Two Harpoon anti-ship missile systems, Sea Sparrow SAM system, two Mk32 double torpedo launchers, one 76mm (3in) Oto Melara gun
Aircraft:	Two NH90 helicopters
Complement:	230 plus 13 aircrew
Launched:	October 2000

Ma'anshan

This 'stealth' frigate carries formidable missile armament and a full range of radar systems, with many features developed from French originals. The sonar is however believed to be a Russian MGK-335 fixed sonar suite. Only two were built before the even more sophisticated *Type 054A* was introduced: an indication of the speed of Chinese naval development.

SPECIFICATIONS

Type:	Chinese Type 054 frigate
Displacement:	4118 tonnes (4053 tons)
Dimensions:	134m x 16m (439ft 7in x 52ft)
Machinery:	Twin screw CODAD drive from four SEMT Pielstick diesel engines, most likely Type 16 PA6 STC
Top speed:	27-30 knots
Main armament:	Two quadruple launchers for VJ83 anti-ship cruise missiles, eight-cell Hong Qi7 short-range SAM system, four 6-barrel 30mm (1.18in) AK-630 CIWS, two Type 87 6-tube ASROC launchers, one 100mm (3.94in) gun
Aircraft:	One Kamov Ka 28 'Helix' or Harbin Z-9C helicopter
Launched:	September 2003

TIMELINE

 2000   2003    2004

Houbei class

Navies are adopting missile-armed fast attack craft, more heavily armed than earlier patrol boats. This class, also known as Type 002, catamaran-hulled with stealth features, is in serial production in China. Detection equipment includes surface search and navigational radars and HEOS300 electro-optics. The two diesel engines generate 5,119 kW (6,865hp). Duties are coastal and inshore patrols.

SPECIFICATIONS	
Type:	Chinese missile boat
Displacement:	223.5 tonnes (220 tons) full load
Dimensions:	42.6m x 12.2m x 1.5m (139ft 7in x 40ft x 4ft 10in)
Machinery:	Twin water jet propulsors, diesels
Top speed:	36 knots
Main armament:	Eight C-801/802/803 anti-ship, or eight Hongniao long-range cruise missiles, twelve QW MANPAD surface-to-air missiles, one 30mm (1.18in) gun
Complement:	12
Launched:	April 2004

Daring

First of eight planned Type 45 destroyers, replacing Type 42, *Daring* was constructed in six 'blocks' in three different yards before final assembly. Its primary purpose is air defence and it has SAMPSON multi-function air tracking and S1850M three-dimensional air surveillance radars. But it is also fitted with MFS-7000 sonar and SSTDS underwater decoy systems.

SPECIFICATIONS	
Type:	British destroyer
Displacement:	8092 tonnes (7962.5 tons) full load
Dimensions:	152.4m x 21.2m x 7.4m (500ft x 69ft 6in x 24ft 4in)
Machinery:	Twin screws, integrated full electric propulsion, gas turbines
Top speed:	29+ knots
Main armament:	SYLVER missile launcher, MBDA Aster 15 and 30 missiles, 2 Phalanx 20mm (0.79in) CIWS
Aircraft:	One Lynx HMA 8 or Merlin HM 1helicopter
Complement:	190
Launched:	February 2006

Aquitaine

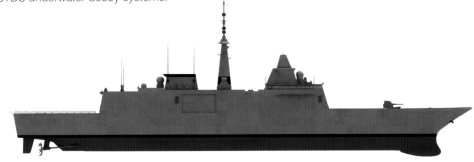

Product of French-Italian co-operation, the FREMM multi-purpose frigate can be adapted to anti-air, anti-ship, and anti-submarine operation. *Aquitaine* and eight sister ships are due for launching from 2012; a further two will be deployed as anti-aircraft ships. Italy plans to build ten; six as general-purpose, four primarily for ASW action. Other NATO navies are likely to acquire them.

SPECIFICATIONS	
Type:	French FREMM-type frigate
Displacement:	6000 tonnes (5544 tons)
Dimensions:	142m x 20m x 5m (465ft 9in x 65ft 6in x 16ft 4in)
Machinery:	Twin screws, integrated full electric propulsion, gas turbines
Top speed:	27+ knots
Main armament:	MM-40 Exocet block 3, MU 90 torpedoes
Aircraft:	One NH90 helicopter
Complement:	108
Launched:	February 2012

2006  2012

Cruise Ships

The popularity of cruising continued into the 2000s and this was reflected by the construction of a new generation of 'super cruise ships', where the ship itself is the resort, fulfilling all needs and making few landfalls. The Gross Tonnage of such vessels, based on total enclosed volume, expressed in tons, reflects their huge enclosed space.

Queen Mary 2

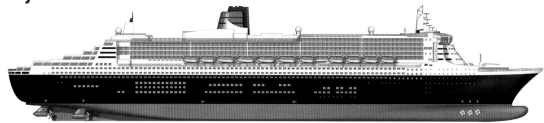

If this is a liner, as owners Cunard claim, it is certainly the largest ever built. Its lines deliberately reflect the Cunarders of old. And it does make scheduled transatlantic runs in summer, with cruising at other times. Its height of 72m (236ft 2in) just gets it under New York's Verrazzano Narrows bridge. QM2 accommodates 3506 passengers.

SPECIFICATIONS	
Type:	British cruise ship
Gross tonnage:	150,904 tonnes (148,528 tons)
Dimensions:	345m x 45m x 10.1m (1132ft x 147ft 6in x 33ft)
Machinery:	Four electric propulsion pods, powered by diesels and gas turbines
Top speed:	29.62 knots
Routes:	Transatlantic and worldwide
Complement:	1,253
Launched:	March 2003

Queen Victoria

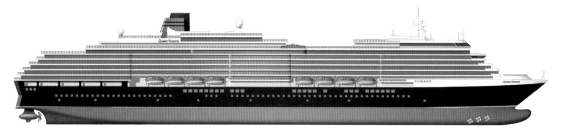

Built by Fincantieri Marghera in Italy for the Cunard line, this ship made its maiden voyage in December 2007. It can take 2014 passengers on short or extended cruises to anywhere in the world, though it cannot traverse the Panama Canal. With 16 decks, of which 12 are for passengers, it has seven restaurants, a ballroom and a theatre.

SPECIFICATIONS	
Type:	British cruise ship
Gross tonnage:	91,440 tonnes (90,000 tons)
Dimensions:	294m x 36.6m x 8m (964ft 6in x 120ft x 26ft 2in)
Machinery:	Two azipod thrusters, diesel-electric
Service speed:	23.7 knots
Routes:	Worldwide
Complement:	2165
Launched:	January 2007

TIMELINE

2003  2007

Poesia

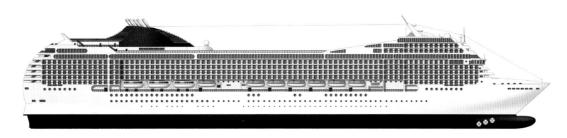

One of the Italian MSC cruise fleet, *Poesia* was built at St Nazaire, France, at a cost of $360 million. Passenger attractions include beauty salon, treatment rooms, gymnasium and jogging track, together with a range of restaurants. lounges and entertainments. Maximum accommodation is for 3605 persons, on 13 decks, with 13 elevators to ease movement.

SPECIFICATIONS

Type:	Panama-registered cruise ship
Gross tonnage:	93,970 tonnes (92,490 tons)
Dimensions:	293.8m x 32.19m x 7.99m (963ft 10in x 105ft 6in x 26ft 2in)
Machinery:	Twin screws, diesels
Service speed:	23 knots
Routes:	Mediterranean
Complement:	987
Launched:	August 2007

Amacello

Said on its launch to inaugurate a new level of comfort in river cruising, *Amacello* holds 148 passengers on 4 decks. Its first cruise was in March 2008. Built for US-owned Amawaterways by Scheepswerf Grave in the Netherlands, it is one of a growing fleet plying not only in Europe, but on the Mekong in Vietnam.

SPECIFICATIONS

Type:	Swiss river cruise-boat
Dimensions:	109.75m x 11.6m x 1m (360ft x 38ft x 3ft 3in)
Machinery:	twin screws, diesels
Routes:	Rivers Rhine, Danube
Complement:	41
Launched:	2007

Oasis of the Seas

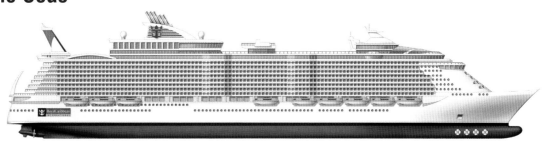

The largest passenger vessel yet built, this ship can take 6296 passengers at maximum occupancy. With a height above the waterline of 72m (236ft 2in), its funnel has retractable flues to enable it to pass under certain bridges. It has no rudder, using its three thruster pods for steering. Based at Port Everglades, Florida, its maiden voyage was made in December 2009.

SPECIFICATIONS

Type:	Bahamas-registered cruise ship
Gross tonnage:	225,282 tonnes (222,238 tons)
Dimensions:	360m x 60.5m x 9.3m (1181ft x 198ft x 31ft)
Machinery:	Three 20MW azimuth thrusters, powered by six diesel engines
Top speed:	22.6 knots
Routes:	Caribbean, Florida coast
Complement:	2165
Launched:	November 2008

2008

Twenty-First Century Cargo Ships

Every commercial ship burns oil, often high-sulphur content 'bunker oil'. Ships are estimated to contribute anything between 1.75 per cent and 4.5 per cent of atmospheric 'greenhouse gas' emissions, and much attention has been focused on ways in which this can be reduced. National and international regulations require ships to contain or process their own waste products.

Berge Bonde

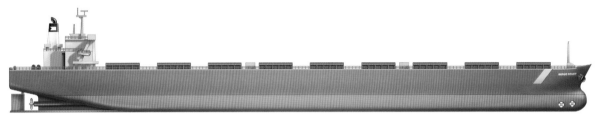

Built by Imabari Shipbuilding, Japan, and owned by La Darien Navigacion, Berge Bond is an ore carrier, with nine holds. Fully loaded, it displaces 206,312 tonnes (203,011 tons) against a net tonnage (measurement of the space available for cargo, expressed in tonnage) of 66,443 tonnes (65,380 tons). In volume terms, its cargo capacity is 220,022m³ (7,766,776 cubic ft).

SPECIFICATIONS

Type:	Panama-registered bulk carrier
Gross tonnage:	104,727 tonnes (103,051 tons)
Dimensions:	299.94m x 50m x 18.1m (982ft x 164ft x 59ft 4in)
Machinery:	Single screw, diesels
Top speed:	15.1 knots
Cargo:	Mineral ore
Routes:	Worldwide
Launched:	2005

Emma Maersk

The world's biggest container ship when launched, *Emma Maersk's* Wärtsila-Sulzer 14RT FLEX96-C diesel engine is also the world's biggest, weighing 2300 tons (2337 tonnes) and generating 109,000hp (82MW). Exhaust gases are used to generate electricity. The sub-waterline hull is painted with a silicone-based paint which reduces drag, deters barnacles and is not inimical to marine life.

SPECIFICATIONS

Type:	Danish container ship
Gross tonnage:	170,974 tonnes (168,239 tons)
Dimensions:	397m x 56m x 15.5m (1300ft x 180ft x 51ft)
Machinery:	Single screw, diesels
Service speed:	25.5 knots
Cargo:	Containers
Routes:	Chinese ports-Algeciras-Rotterdam-Bremerhaven
Complement:	13
Launched:	2006

TIMELINE   2005   2006   2008

Grand Victory

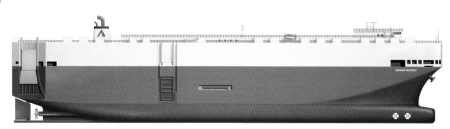

While it is easy to feel all car carriers look the same, *Grand Victory* was designed with an innovatory, asymmetric stern fin and an unusual form of bulbous bow, aimed at greater manoeuvrability. Its engine was also designed to minimise emissions, noise and vibration. A ro-ro ship, with two boarding ramps, it carries cars on 12 decks.

SPECIFICATIONS

Type:	Japanese car carrier
Gross tonnage:	60,164 tonnes (59,217 tons)
Dimensions:	199.99m x 35.8m x 9.62m (656ft x 117ft 5in x 31ft 8in)
Machinery:	Single screw, diesel
Top speed:	19.8 knots
Cargo:	6402 cars
Routes:	Japan-Asian and American ports
Launched:	June 2008

Auriga Leader

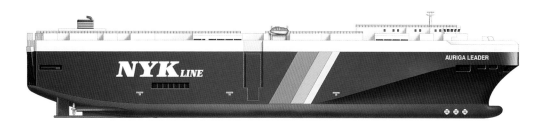

Hailed as 'ship of the year' by Lloyd's List in 2009, *Auriga Leader's* deck is covered by 328 solar panels. The power they produce makes a very modest contribution to the ship's energy requirement, only 0.05 per cent of propelling power, but it is an experimental and unique vessel in this respect and further development is hoped for.

SPECIFICATIONS

Type:	Japanese car carrier
Gross tonnage:	61,176 tonnes (60,213 tons)
Dimensions:	199.99m x 32.26m x 9.7m (656ft 1in x 105ft 9in x 31ft 10in)
Machinery:	Single screw, part solar-powered, diesels
Cargo:	6200 cars
Routes:	Japan-California
Launched:	2008

Cheikh el Mokrani

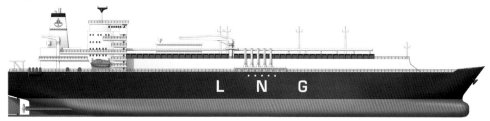

A member of the Mediterranean LNG Transport Corporation fleet, this ship shuttles across the Mediterranean Sea to maintain gas supplies in southern European countries. Total capacity is 75,500m³ (2,665,150 cubic ft) and the cargo tanks are finished and installed in line with the latest requirements of IMO and MARPOL. A bow thruster helps work the ship in harbours.

SPECIFICATIONS

Type:	Bahamas-registered liquefied gas tanker
Gross tonnage:	52,855 tonnes (52,009 tons)
Dimensions:	219.95m x 22.55m x 9.75m (721ft 5in x 74ft x 32ft)
Machinery:	Single screw, steam turbines, bow thruster
Top speed:	17.5 knots
Cargo:	Liquefied natural gas
Routes:	Algerian ports to Spanish, French and Italian ports
Launched:	2008

Specialized Vessels

Specialized ships have become even more specialised, sophisticated and capable in recent years. One of the key new technologies has been that of 'dynamic positioning', enabling a vessel to maintain a very precise station or line. In this field, even more than in general shipping, environmental impact is a major factor and is closely monitored.

Blue Marlin

Specifications are of the reconstructed *Blue Marlin* (2004). Originally Norwegian-owned, the vessel was acquired by Dockwise Shipping of the Netherlands in 2001. It carries structures of up to 60,000 tonnes and can be submerged by 10m (33ft) for them to be floated on board. Major tasks have included carrying the 60,000 tonne oil platform 'Thunder Horse.'

SPECIFICATIONS

Type:	Dutch heavy lift ship
Displacement:	76,060 tonnes (74,843 tons) maximum load
Dimensions:	224.5m x 42m x 13.3m (712ft x 138ft x 44ft)
Machinery:	Single screw, diesels, two retractable peropulsors
Top speed:	14.5 knots
Complement:	60
Launched:	April 2000 (rebuilt 2004)

HAM 318

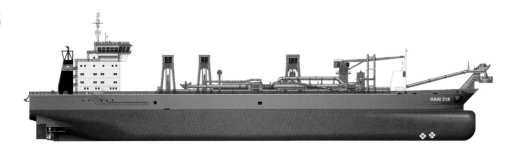

With two 12m (39ft 4in) diameter suction pipes, this hopper dredger can operate to a depth of 17m (55ft), discharging sludge through bottom doors or land-pipe. It can be used in clearing oil spillages and has storage for 2,800m³ (98,840 cubic ft) of heavy oil. It is planned to increase its dredging depth to 110m (360ft).

SPECIFICATIONS

Type:	Dutch dredger
Displacement:	57,360 tonnes (56,442 tons)
Dimensions:	176.15m x 32m x 13m (577ft 9in x 105ft x 42ft 6in)
Machinery:	Twin controllable-pitch screws, diesels, two bow thrusters
Top speed:	17.3 knots
Complement:	45
Launched:	October 2001

TIMELINE

2000  2001

Tyco Reliance

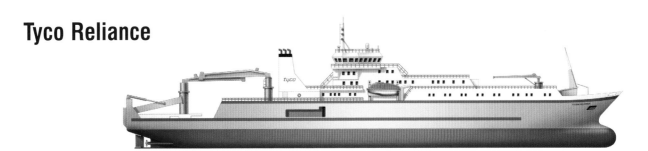

Built by Keppel Singmarine of Singapore, this deep sea cable layer is operated by Tyco Telecommunications. It is capable of carrying 5465 tonnes (5377.5 tons) of cable, and is equipped with Kongsberg Simrad SDP 21 dynamic positioning system. It carries a Perry Tritech ST200 remotely operated vehicle (ROV) for seabed work, and a trenching sea-plough.

SPECIFICATIONS	
Type:	Marshall Islands registered cable layer
Displacement:	12,184 tonnes (11,989 tons) full load
Dimensions:	140m x 21m x 8.4m (459ft 2in x 68ft 11in x 27ft 6in)
Machinery:	Two azimuth propulsors, bow and stern thrusters, diesel electric
Top speed:	13 knots
Complement:	80
Launched:	2001

Stena Drill MAX

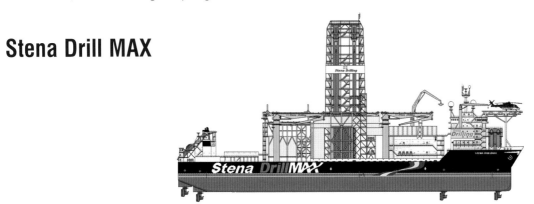

Built in South Korea, capable of working in waters over 3,000m (10,000ft) deep and of drilling to a depth of 10,670m (35,000ft), in sub-Arctic conditions, this vessel uses dynamic positioning systems and multiple thrusters to remain on station. The deck can bear loads up to 15,000 tonnes (14,760 tons). A helicopter deck is fitted.

SPECIFICATIONS	
Type:	Swedish deep sea drilling ship
Displacement:	96,000 tonnes (94,464 tons)
Dimensions:	228m x 42m x 19m (748ft x 137ft 10in x 62ft 4in)
Machinery:	Six sets of azimuth thrusters, diesel electric
Top speed:	12 knots
Complement:	180
Launched:	2007

Crestway

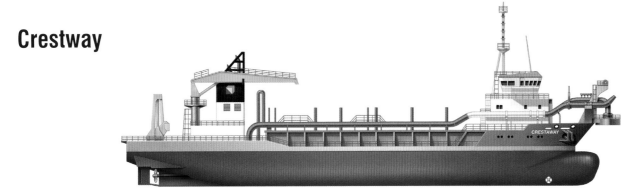

Operated by the Belgian dredging company Royal Boskalis, *Crestway* is a suction dredger able to work in depths of up to 33m (110ft) and with a hopper capacity of 5600m^3. Ships like this are a major investment and can operate virtually anywhere, on channel deepening, foreshore restructuring, and similar large-scale tasks.

SPECIFICATIONS	
Type:	Cyprus-registered dredger
Gross tonnage:	5005 tonnes (4925 tons)
Dimensions:	97.5m x 21.6m x 7.6m (319ft 10in x 70ft 10in x 23ft 7in)
Machinery:	Two azimuth propulsors, bow and stern thrusters, diesel electric
Top speed:	13 knots
Complement:	14
Launched:	May 2008

2007   2008

Glossary

AA Anti-aircraft, as in 'anti-aircraft artillery' (AAA); air-to-air, as in 'air-to-air missile' (AAM).

Armour Plates of iron, later steel and alloys, later still more exotic metals such as titanium, added to the hull and essential components of a ship to protect its integrity, its vital parts and its crew from battle damage.

ASM Anti-submarine missile; anti-submarine mortar (also air-to-surface missile, air-to-ship/anti-ship missile).

ASW Anti-submarine warfare.

Axial fire Gunfire ahead or astern, along the major axis of the vessel.

Ballast The weight added to a ship or boat to bring her to the desired level of floatation and to increase stability.

Barge Most commonly, a flat-bottomed vessel of shallow draught used to carry cargo on inland waterways, both from port to port and to and from ocean-going ships.

Battlecruiser The made-up designation for a hybrid warship armed like a battleship but sacrificing passive protection in the form of armour plate for speed.

Battleship Originally the biggest and most powerful ships of the fleet, mounting guns of usually 10in (254mm) or larger calibre (the biggest were those of the Japanese Yamato class, which were 18.1in (460mm)), and heavily armoured.

Beam The width of a ship's hull.

Beam engine The original form of single-cylinder steam engine, the piston of which acted on a beam by way of a connecting-rod. The beam itself formed a simple Class I lever, its reciprocating action being translated into rotation by a link to a crank or eccentrically to a flywheel.

Bofors A Swedish armaments manufacturer, best known for its 40mm anti-aircraft gun.

Boiler A device for heating water to boiling point, to turn it into steam so that it could be employed to power machinery.

Bonaventure An extra mizzen sail; it fell out of use towards the end of the 17th century.

Boom A spar used to extend the foot of a sail; also a floating barrier, usually across the entrance to a harbour.

Bore The diameter of a cylinder or gun barrel.

Bowsprit A spar protruding over the bows of a sailing vessel, to serve as an outboard anchorage for the tack of a flying jib.

Break A change in the level of a deck, eg 'the break of the forecastle'.

Breech block The removeable part of a gun's breech, through which projectile and charge could be loaded.

Breech-loading (BL) Guns loading from the breech, via a removable segment.

Broadside The side of the ship; the simultaneous firing of the guns located there.

Bulges/Blisters Chambers added to the outside of a warship's hull to provide protection against torpedoes; outer sections were generally water-filled, to absorbe splinters, inner, air-filled, to diffuse blast.

Bulk Carrier A single-deck ship expressly designed to accommodate loose cargo such as grain or ore.

Bulkhead/Water-tight bulkhead A vertical partition employed to divide up a ship's internal space, both longitudinally and transversely. These partitions may be water-tight, in which case the openings in them to allow passage must be capable of being sealed, preferably by remote control.

Bunker/Bunkerage The part of a ship allocated to the storage of fuel; the fuel itself, and the quantity carried.

Calibre the diameter of the bore of a gun barrel; the number of times that diameter fits into the length of the barrel, expressed as 'L/(calibre)'; eg, a gun of 10in bore with a barrel 300in long would be described as '10in L/30', or just '10in/30'.

CAM ship Catapult-Armed Merchant ship A merchant ship equipped to fly off a fighter aircraft; a temporary expedient adopted during 1940.

Capital Ship A term coined around 1910 to describe the most important naval assets, and group together battleships and battlecruisers (chiefly to give extra credibility to the latter); it was later extended to include monitors.

Carrier battle group A force designation coined during World War II; it was made up of one or more fleet aircraft carriers together with associated defensive elements – destroyers and cruisers – but often included battleships, which had by then largely been relegated to the shore bombardment role.

Carronade A short-barrelled, lightweight muzzle-loading gun, produced by the Carron Ironworks in Scotland from the 1770s.

Casemate A fixed armoured box within which a gun was mounted. It allowed the weapon to be elevated and trained, and usually protruded from the hull or superstructure to increase its arc of fire.

Catamaran A boat (and later ship) with two hulls joined by a continuous deck or decks.

cb cabin class

Citadel/Central citadel A heavily-armoured redoubt within which the ship's main battery was housed.

Clipper An ultimately meaningless term used to describe any fast sailing ship, particularly one engaged in the grain, opium or tea trades, widely used in the mid-19th century.

Combined Carrier A cargo ship of the latter part of the 20th century, adapted to carry both bulk and containerised cargoes.

Composite construction A construction method employing iron or steel frames and wooden hull planking.

Compound engine A multi-cylinder steam engine in which the steam is employed at least twice, at decreasing pressures; theoretically, all multi-cylinder engines are compound engines, but those which employ the steam three times are known as triple-expansion, and those which employ it four times as quadruple-expansion engines.

Conbulker *see Combined Carrier.*

Corvette Originally a (French) sailing ship of war, too small to warrant a rate (and thus the equivalent of the British sloop); more recently, a warship smaller than a frigate or destroyer-escort.

Cruiser A warship, larger than a frigate or destroyer, much more heavily armed and often armoured to some degree, intended for independent action or to act as a scout for the battlefleet. Modern cruisers operate as defensive elements within carrier battle groups.

Cutter Originally a small, decked boat, lightly-armed, with a single mast and bowsprit, carrying a fore-and-aft (gaff) mainsail and a square topsail, with either two jibs or a jib and staysail, frequently used as an auxiliary to the fleet or on preventive duties.

CVA Attack aircraft carrier.

DDG Guided-missile destroyer.

DDH Helicopter destroyer.

Deadrise The angle to the vertical of the planking in the floor of a vessel's hull, and thus a measure of the 'sharpness' of her lines.

Deadweight *see Tonnage.*

Deck The continuous horizontal platforms, the equivalent of floors in a building, which separate a ship.

Derrick A form of lifting gear comprising a single spar attached at its heel low down to a mast or kingpost and pivoted there, equipped with stays, a topping lift and guy pendants, so that its attitude and position may be positively controlled, and with a block attached at its head, through which a runner may be rove and led, if necessary, to a winch.

Destroyer Originally torpedo-boat destroyer; a small warship of little more than 200 tons displacement, itself equipped to launch torpedoes, but also armed with light guns.

Destroyer-escort A small warship, bigger than a corvette of the period, designed and constructed to guard convoys of merchant ships, especially on the Atlantic routes, during World War II.

Diesel A form of internal-combustion engine which ignites its fuel/air mixture by compression alone, invented by Rudolf Diesel in 1892.

Diesel-electric A form of propulsion in which the compression-ignition internal-combustion engine drives a generator, which in turn drives an electric motor which turns the propeller shaft.

Displacement A measure of the actual total weight of a vessel and all she contains obtained by calculating the volume of water she displaces.

Draught (also Draft) The measure of the depth of water required to float a ship, or how much she 'draws'.

Dreadnought The generic name given to a battleship modelled after HMS Dreadnought, the first with all-big-gun armament; it fell into disuse once all capital ships were of this form.

Driver An additional sail hoisted on the mizzen; see Spanker.

ECM Electronic Countermeasures; measures taken to decoy or confuse an enemy force's sensors.

EEZ Economic Exclusion Zone; a coastal strip, 200nm wide measured from the mean low water mark, within which a nation is held to have the exclusive right to exploit natural resources; it supplements, but does not replace, the territorial limit (originally three miles, the effective range of coastal artillery, later extended to 12 miles).

Fast attack craft Gun, torpedo and/or guided-missile-armed warships, characterised by their small size and high speed. Such craft only ever had limited success.

Fire control A (centralised) system of directing the firing of a ship's guns, based on observation of the fall of shot relative to the target, taking into account movements of both the target and the firing platform.

Flag of Convenience *see Registry.*

Flare The outward (usually concave) curve of the hull of a ship towards the bow.

Floflo Float on/float off loading operations; *see Lolo; Roro.*

Flotilla In the Royal Navy up to World War II, an organised unit of (usually eight) smaller warships – destroyers and submarines in particular, but also minesweepers and fast attack craft; cruisers and capital ships were grouped into squadrons, and squadrons and flotillas made up fleets – derived from the diminutive of the Spanish flota, fleet.

Flush decked Commonly, a ship with no forecastle or poop.

Forced draught A means of increasing the efficiency of a ship's boilers by forcing air at higher than normal pressure through the furnace element.

Fore-and-aft Sails which, when at rest, lie along the longitudinal axis of a vessel; a vessel rigged with such sails.

Forecastle Originally the superstructure erected at the bows of a ship to serve as a fighting platform, later the (raised) forward portion and the space beneath it, customarily used as crews' living quarters. Pronounced fo'c'sle.

Frames The ribs of a vessel, upon which the hull planking is secured, set at right-angles to the keel.

Freeboard The distance between the surface of the water and the upper deck of a ship or the gunwale of a boat.

Frigate Originally, fifth- or sixth rate ships carrying their guns on a single deck, employed as scouts, and the counterpart of the later cruiser.

Gas turbine A rotary internal-combustion engine in which a fuel/air mixture is burned and the rapidly-expanding gas thus produced used to drive turbine blades arranged upon a shaft; a gas turbine bears the same relationship to a steam turbine that a reciprocating internal-combustion engine does to a reciprocating steam engine.

Gundeck The name given to the main deck in sailing warships of the Royal Navy.

Head The bows of a ship (and as in '(down) by the head'), and by extension in the plural, because they were traditionally located there, the latrines; also the top of a four-sided sail.

Headsail Those sails – jibs and staysails in the main, but also spinnakers – which are set before the (fore)mast.

Heel, to The act of temporarily tipping a vessel to one side, usually caused by the pressure of wind on sails.

Heeling tanks Ballast tanks set on a ship's sides to allow her to be deliberately heeled over.

Horsepower A measure of the power produced by an engine; one horsepower = 550 foot/pounds per second ('the power required to raise 550 pounds through one foot in one second') as defined by James Watt.

Ironclad The contemporary name for wooden warships clad with iron, and by extension, to the first iron warships; it continued in use up until the arrival of the dreadnought.

Jib A triangular sail (usually loose-footed) set on a forestay.

Jib-boom A continuation of the bowsprit.

Keel The main longitudinal timber of a ship or boat, effectively her spine and certainly her strongest member. In yachts, a downwards extension in the form of a wing (or wings) which balances the pressure of the wind upon the sails; also, the flat-bottom lighters used in ports in the northeast of England.

Ketch A two-masted sailing vessel; in modern terms, a yacht with a reduced mizzen stepped before the rudder post (see yawl).

Knot Internationally, the measure of a ship's speed – one nautical mile per hour.

Leeboard A primative form of drop-keel, hung over the side of a vessel from a forward pivot. Two were carried, one to port and one to starboard, though only the one on the lee side was employed.

Lifting screw A screw propeller designed to be lifted clear of the water, fitted to sailing ships with auxiliary machinery, the object of the exercise being to reduce drag when sailing.

Lighter A 'dumb' (ie, unpowered) barge, used as a transit vehicle to load and unload ships in port and also in places where proper port facilities are deficient.

Liner A ship carrying passengers to a fixed schedule, usually on trans-oceanic routes; the term became current from the mid-1800s. A cargo liner also operates on fixed schedules, with space for a limited number of passengers.

Magazine Secure storage for explosives.

Masts Spars, mounted vertically or close to it, normally stayed (guyed) at either side and fore and aft, employed to allow sails to be carried.

MGB/MTB Motor gun boat/motor torpedo boat; *see Fast attack craft.*

Minesweeper A small ship, roughly the size of a trawler (many were, in fact, converted fishing boats originally) adapted and equipped to locate and neutralise submarine mines. Later supplemented by specialist minehunters.

Mizzen The third mast, counting from the bows; since most ships had three masts, it was also the aftermost, and invariably carried a fore-and-aft steadying sail.

NATO North Atlantic Treaty Organisation.

Nautical mile Internationally, the measure of distance at sea which has become standardised at 6080ft (1852m).

OBO Ore/Bulk/Oil carrier.

Oerlikon A Swiss arms manufacturer whose 20mm cannon, widely acknowledged to be the best of its type, was adopted by both sides during World War II. Improved versions were still in production at the end of the 20th century.

Periscope An optical device allowing an observer to change his plane of vision. At sea they were commonly used to allow submerged submarines a view of the surface, and also in gun turrets.

Poop The short raised deck at the stern of a vessel, originally known as the aftercastle.

Propeller Properly speaking, the screw propeller; as essential to steam- and motor ships as their powerplant, the rotation of the screw propeller and the angle of its blades or vanes combine to generate thrust against the mass of water, which pushes the vessel through it.

Propulser Commonly, any propulsive device – a water jet, for example – which is not a propeller; the most effective is probably the Voith-Schneider.

Quarterdeck That part of the upper deck abaft the mainmast (or where the mainmast would logically be in a steam- or motor ship), traditionally the reserve of commissioned officers.

Quick-firing A designation applied to small- and medium-calibre guns to indicate that they used unitary ammunition (ie, with projectile and propellant cartridge combined).

Radar An acronym for Radio Direction and Range – a means of using electromagnetic radiation to locate an object.

Reserve Warships not in active commission are said to be in reserve; this may be a temporary measure, in which case maintenance work will be kept fully up to date, or a long-term measure, in which case the ship will be 'mothballed'.

Rudder A vertical board or fin hung on the centreline of the vessel at the stern post, originally (and still, in small boats) from simple hinges known as pintels, and connected either directly to the tiller or by ropes or chains to the steering wheel, which, when it is angled relative to the vessel's course, causes a change of direction.

SAM Surface-to-air missile.

Schooner-rigged A boat or ship with two or more masts of equal height (or with the foremast lower than the main and others), fore and aft rigged on all of them, with or without topsails.

Schnorkel/Snorkel A tube with a ball-valve at its upper extremity, which allows a submarine to take in air, and thus continue to operate its internal-combustion engines, while remaining below the surface.

Seabee A system of carrying loaded cargo barges aboard ocean-going ships to simplify on- and off-loading operations at terminal ports. Barges (more properly, lighters) were lifted on and off by means of an elevator at the stern of the ship; see LASH.

Sheer The upward curve of a ship's upper deck towards bow and stern.

Smooth-bore A gun with a smooth (ie, unrifled) barrel, used as naval ordnance until the second half of the 19th century.

Sonar An acronym for Sound Navigation and Ranging, a technique of using sound waves to detect objects underwater, and by extension, to the hardware employed; see also ASDIC.

Spanker Originally an additional sail hoisted on the mizzen mast to take advantage of a following wind, later taking the place of the mizzen course; see Driver.

Spar deck Strictly speaking, a temporary deck, but later used to describe the upper deck of a flush-decked ship.

Sponson A platform built outside the hull, at main- or upper deck level, usually to allow guns on the broadside to be sited so as to allow them to fire axially.

Squadron In the Royal Navy, originally an organised unit of (usually eight) major warships – cruisers and capital ships, but in the US Navy (and the practice became widespread), an organised unit of ships of any type, from minesweepers upwards, the term having taken over from flotilla.

Square-rigged A sailing vessel whose sails are set on yards, which when at rest are at right-angles to the longitudinal axis of the hull.

SSM Surface-to-surface missile.

SSN Nuclear-powered submarine.

Standing Rigging That portion of a ship's rigging – stays and shrouds, for example – which is employed to steady her masts; see also Running rigging.

Steam Turbine A rotary engine in which steam is used to drive turbine blades arranged upon a shaft.

Stem The foremost member of a ship's frame, fixed at its lower extremity to the keel.

Stern post The aftermost member of a ship's frame, fixed at its lower extremity to the keel.

Submarine A vessel capable of indefinite (or at least very prolonged) underwater operation.

Tack The lower forward corner of a fore-and-aft sail; a reach sailed (in a sailing vessel) with the wind kept on one side.

Tiller A wooden or metal bar attached rigidly to the rudder and used to control its movement.

Tonnage The load carrying capacity of a merchant ship or the displacement of a warship.

Topmast The second section of a mast, stepped above the lower mast, carrying the (upper and lower) topmast yard(s).

Topsail In a square-rigged vessel, the square sails set immediately above the course, from the topmast yard(s) (bigger ships carried paired topsails, upper and lower, for ease of working); in a fore-and-aft rigged vessel, the topsails may be either square or themselves fore-and-aft.

Torpedo A self-propelled explosive device, with or without some form of guidance, running on or below the surface of the sea.

Trawler A fishing boat (in fact, the largest are substantial ships) which drags behind it on two warps a roughly cone-shaped net. Modern trawlers have substantial refrigeration plant and freezers, and stay at sea for weeks at a time.

Triple-expansion A type of reciprocating compound engine, with a minimum of three cylinders of graduated sizes housing pistons connected to a common crankshaft; the steam, introduced at very high pressure into the smallest, is condensed, and passes to the second, slightly larger, cylinder, where it is condensed once more, and then passed to the largest cylinder.

Turbo-electric A form of propulsion in which the steam turbine drives a generator, which in turn drives an electric motor which turns the propeller shaft.

Turret Originally an armoured shell or covering for a gun, which rotated with the platform upon which the gun is mounted; later the armoured cover became an integral part of the rotating mounting, and itself supported the gun or guns.

Turret ship A design of cargo carrier developed in the 1890s, its upper deck being about half the full beam of the ship; slightly below this was the so-called 'harbour deck', which joined the main vertical plating in a wide-radius curve. It was basically a subterfuge, which came about as a result of the Suez Canal Company's policy of charging dues based on the ship's beam at the upper deck.

Yawl A two-masted sailing yacht with a reduced mizzen stepped abaft the rudder post (see ketch).

Ships Index

General Index

Picture Credits